Weed Science

Weed Science

Clara Fanning

SYRAWOOD
PUBLISHING HOUSE

New York

Published by Syrawood Publishing House,
750 Third Avenue, 9th Floor,
New York, NY 10017, USA
www.syrawoodpublishinghouse.com

Weed Science
Clara Fanning

International Standard Book Number: 978-1-64740-070-5 (Hardback)

This book contains information obtained from authentic and highly regarded sources. All chapters are published with permission under the Creative Commons Attribution Share Alike License or equivalent. A wide variety of references are listed. Permissions and sources are indicated; for detailed attributions, please refer to the permissions page. Reasonable efforts have been made to publish reliable data and information, but the authors, editors and publisher cannot assume any responsibility for the validity of all materials or the consequences of their use.

Trademark Notice: Registered trademark of products or corporate names are used only for explanation and identification without intent to infringe.

Cataloging-in-Publication Data

Weed science / Clara Fanning.
     p. cm.
Includes bibliographical references and index.
ISBN 978-1-64740-070-5
1. Weeds--Control. 2. Weeds. 3. Herbicides. I. Fanning, Clara.
SB611 .W44 2022
632.5--dc23

TABLE OF CONTENTS

This book is a culmination of my many years of practice in this field. I attribute the success of this book to my support group. I would like to thank my parents who have showered me with unconditional love and support and my peers and professors for their constant guidance.

The plants that are considered as undesirable in the farms, lawns, gardens, parks and fields are known as weeds. The discipline of weed science focuses on weeds, their effect upon human activities as well as their management. It is a branch of applied ecology which seeks to modify the environment against natural evolutionary trends. Various methods such as powered cultivation with cultivators, cultivation with hoes and smothering with mulch and soil solarisation, chemical attack with herbicides, lethal wilting with high heat and burning are used to control weed. This book provides comprehensive insights into the field of weed science. While understanding the long-term perspectives of the topics, it makes an effort in highlighting their impact as a modern tool for the growth of the discipline. This textbook is appropriate for those seeking detailed information in this area.

The details of chapters are provided below for a progressive learning:

Chapter – Introduction

Weed refers to the unwanted wild plants that grow in gardens and fields. Weed science deals with the study of management of weeds in agricultural production systems, natural areas, and managed properties in residential and urban areas. This is an introductory chapter which will briefly introduce about weed science.

Chapter – Types of Weeds

Some of the important classifications of weed include buddleja, cirsium arvense, knapweeds, ragweed, euphorbia lathyris, equisetum, amaranth, chenopodium album, glechoma hederacea, jacobaea vulgaris, etc. The topics elaborated in this chapter will help in gaining a better perspective about these types of weeds.

Chapter – Invasive Plants

Invasive plants are the species of plants that are not native to a specific location and can pose serious environmental threats. Tamarix, eichhornia crassipes, reynoutria japonica, cynodon dactylon, ailanthus altissima, cyperus esculentus, purple loosestrife, impatiens glandulifera, etc. are some of the invasive plants. This chapter closely examines these types of invasive plants to provide an extensive understanding of the subject.

Chapter – Uses of Weeds

Weeds are beneficial and have a number of applications in the environment. It includes fertilizing the soil, increasing moisture, acting as shelter, repelling pests, attracting beneficial insects, serving as food to humans, etc. All these diverse uses of weeds have been carefully analyzed in this chapter.

Chapter – Weed Ecology and Biology

Weed ecology studies the interaction between weeds and its environment. The study of the growth, reproduction and life cycles of weed species and its vegetation is called weed biology. This chapter has been carefully written to provide an easy understanding of weed ecology and biology.

Chapter – Methods of Weed Control

Weed control is the method which controls and manages the growth of noxious and invasive weeds. Aquatic weed harvester, stale seed bed, soil steam sterilization, etc. are some of the tools and techniques used for weed control. This chapter discusses about these different tools and related methods of weed control in detail.

Clara Fanning

Introduction

Weed refers to the unwanted wild plants that grow in gardens and fields. Weed science deals with the study of management of weeds in agricultural production systems, natural areas, and managed properties in residential and urban areas. This is an introductory chapter which will briefly introduce about weed science.

WEEDS

Weed is a general term for any plant growing where it is not wanted. Ever since humans first attempted the cultivation of plants, they have had to fight the invasion by weeds into areas chosen for crops. Some unwanted plants later were found to have virtues not originally suspected and so were removed from the category of weeds and taken under cultivation. Other cultivated plants, when transplanted to new climates, escaped cultivation and became weeds or invasive species. The category of weeds thus is ever changing, and the term is a relative one.

Weeds interfere with a variety of human activities, and many methods have been developed to suppress or eliminate them. These methods vary with the nature of the weed itself, the means at hand for disposal, and the relation of the method to the environment. Usually for financial and ecological reasons, methods used on a golf course or a public park cannot be applied on rangeland or in the forest. Herbicide chemicals sprayed on a roadside to eliminate unsightly weeds that constitute a fire or traffic hazard are not proper for use on cropland. Mulching, which is used to suppress weeds in a home garden, is not feasible on large farms. Weed control, in any event, has become a highly specialized activity. Universities and agricultural colleges teach courses in weed control, and industry provides the necessary technology. In agriculture, weed control is essential for maintaining high levels of crop production.

The many reasons for controlling weeds become more complex with the increasing development of technology. Plants become weeds as a function of time and place. Tall weeds on roadsides presumably were not problematic prior to the invention of the automobile. However, with cars and increasing numbers of drivers on roads, tall weeds became dangerous, potentially obscuring drivers' visibility, particularly at intersections. Sharp-edged grasses are nominal nuisances in a cow pasture; when the area is converted to a golf course or a public park, they become an actual nuisance. Poison oak (*Toxicodendron diversilobum*) is rather a pleasant shrub on a sunny hillside in the open country; in a camp ground it is a definite health hazard. Such examples could be given ad infinitum to cover every aspect of agriculture, forestry, highway, waterway and public land management, arboretum, park and golf-course care, and home landscape maintenance.

Weeds compete with crop plants for water, light, and nutrients. Weeds of rangelands and pastures may be unpalatable to animals, or even poisonous; they may cause injuries, as with lodging of foxtails (Alopecurus species) in horses' mouths; they may lower values of animal products, as in the cases of cockleburs (Xanthium species) in wool; they may add to the burden of animal care, as when horses graze in sticky tarweeds (Madia species). Many weeds are hosts of plant disease organisms. Examples are prickly lettuce (Lactuca scariola) and sow thistle (Sonchus species) that serve as hosts for downy mildew; wild mustards (Brassica species) that host clubroot of cabbage; and saltbrush (Atriplex species) and Russian thistle, in which curly top virus overwinters, to be carried to sugar beets by leafhoppers. Many weeds are hosts of insect pests, and a number are invasive species.

SUPERWEEDS

Superweeds are weeds that have built up a resistance to the effects of herbicides used in agriculture to kill them. Weeds such as marestail (horseweed) and palmer amaranth (pigweed) are rapidly growing problems for farmers, pushing them to apply more herbicides, like 2, 4-D and Roundup, more often. These products have been shown to harm human health and the environment and their increased use is a step in the wrong direction.

Superweeds also pose a threat to the overall biodiversity of our landscape. When we increase the use of herbicides that have a limited effect on the weeds targeted, we inadvertently decrease the populations of other natural flora and fauna that provide great benefit to the environment, such as Milkweed (an essential food source for the monarch butterfly) and the northern leopard frog.

The development of superweeds is directly related to the planting of herbicide resistant genetically engineered (GE) crops such as Roundup Ready Corn and Soybeans. They allow for the application of high concentrations of powerful herbicides. Genetically engineered crops were initially created with the goal of reducing the use of herbicides. However, over the years, they have had the opposite effect. The use of glyphosphate, the active ingredient in Roundup, has increased by 26% (81.2 million pounds) between 2001 and 2010.

The use of GE crops resistant to herbicides has created a vicious cycle – the over application of herbicides, leading to the evolution of herbicide resistant superweeds, leading to even more application of herbicides. This needs to stop and it's time to take action to protect the well-being of our communities and the environment.

WEED SCIENCE

Weed Science deals with a serious biotic threat capable of causing heavy economic loss to the farmer: the weeds. The nature of weeds and how they interact with human activities form the basis of the discipline of Weed Science. This topic can be considered as a model to follow in the integration of numerous disciplines, using a systematic approach to solve practical problems. The areas of study range from ecological to agronomical investigation to the design of practical methods for managing weeds in the environment. Weed scientists deal with a wide range of scientific issues with the common aim to find practices necessary for an effective and responsible management of weed species. Weed Science has been highly successful in providing efficient, relatively cheap and safe technology to control weeds in a large variety of crops. Although much of this success is due to the low cost and extremely efficient herbicides, a variety of physical, cultural and biological technologies have also been developed and introduced commercially, as a result of co-evolution with society's demand of more environmentally sound crop management systems. The main goal of weed management is to provide the most appropriate methods to ensure a sustainable ecosystem and minimum influence of nuisance plants in various situations. The primary objective of weed research should be to improve our understanding of the relationship between weeds and crops, with the aim of improving the management and control of weeds. Although weeds are the cause of the highest potential crop losses (-34%), nowadays weeds are frequently underestimated since more attention is paid to insect pests (18% loss) or pathogens (16% loss). The methods used for managing weeds vary, depending on the situation, the available research information, the tools, the economics, and the farmers' experience. Weed management is a process of reducing weed growth to an acceptable level. Weed management has been defined as "rational deployment of all available technology to provide systematic management of weed problems in all situations". Based on this definition, weed management could be considered a systematic approach for minimizing the effects of the weeds and optimizing land use, combining prevention and control. The growing demand for food has gradually changed agricultural practices from intense manual work to methods involving less manual labor and more fossil fuel. Although it is not clear when Weed Science began, over the last century it has certainly become an important issue concerning the study of weeds and their management.

Types of Weeds

2

Some of the important classifications of weed include buddleja, cirsium arvense, knapweeds, ragweed, euphorbia lathyris, equisetum, amaranth, chenopodium album, glechoma hederacea, jacobaea vulgaris, etc. The topics elaborated in this chapter will help in gaining a better perspective about these types of weeds.

CIRSIUM VULGARE

Cirsium vulgare is mostly regarded as a weed or undesirable plant. However, the plants are very attractive and offer lots of rewards for butterflies and other insects, and despite the very sharp thorns on the leaves, it also has some uses for humans.

A spiny, herbaceous, biennial plant, with a large basal rosette of leaves, a deep taproot and a flowering stem growing from the centre of the rosette, up to 1.5 m high. The stems have spiny wings and branch in the upper parts. Leaves are dark green, with stiff hairs above, white woolly below, deeply lobed, with the lobes ending in strong spines.

Young flower heads are teardrop-shaped, with 10 to 12 series of green, spiny bracts, curling outwards. The flowers are all tubular, purplish pink.

Achenes are about 4 mm long and 1.5 mm wide, with many rows of silky-plumed pappus bristles, about 20 mm long. The bristles are often interlocked and the pappus and seed are often distributed in a cluster.

Plants flower during summer, from September to April.

Distribution and Habitat

Cirsium vulgare is distributed throughout the country. It is a ruderal species, which are the first to colonize disturbed areas. It is found in grassland or forested areas, overgrazed lands, pan and dam edges and river banks, vacant fields, burnt areas, roadsides, construction sites or any disturbed sites, in moist, as well as dry conditions.

Ecology

This plant is undesirable and unpalatable to most grazing animals. It is, therefore, very visible in heavily grazed fields because it may be the only plant remaining.

The flowers of *Cirsium vulgare* are a rich source of nectar and pollen. It is utilized by various insects, such as beetles, honey bees, bumble bees, hoverflies and many butterflies.

Seeds are dispersed by wind, animals, mud, water, and sometimes by ants. Human activities such as bailing or harvesting can also aid the dispersal of seeds.

Uses

Although the leaves, stems and roots are edible and can be used cooked as a vegetable or fresh in salads, it is almost not worth the effort. The spines need to be removed, and after this tedious process very little remains to be used. The flowers resemble tiny artichokes, but are even more of an effort to clean and then produce a very bland heart.

A mixture of the soaked leaves and roots of the *Cirsium vulgare* is used as a medicine to heal a stiff neck, seizures, as well as nervous disorders.

The boiled leaves can act as an effective diuretic and mildly drains the liver, as well as provides comfort when fevers are caused by an overworked liver.

The boiled mixture of the whole plant is used traditionally to treat rheumatic joint pains and also to heal bleeding piles. The boiled roots mixture of the plant is used as a poultice to treat aching jaws.

The roots are also reported to be effective in lowering blood sugar and cholesterol levels, and also reduce blood pressure. Roots are also used for muscular inflammation.

The oil from the seeds extracted by means of cold press, is used for cooking and as a lamp oil.

BUDDLEJA

Buddleja, or *Buddleia*, commonly known as the butterfly bush, is a genus comprising over 140 species of flowering plants endemic to Asia, Africa, and the Americas. The generic name bestowed by Linnaeus posthumously honoured the Reverend Adam Buddle (1662–1715), an English botanist and rector, at the suggestion of Dr. William Houstoun. Houstoun sent the first plants to become known to science as buddleja (*B. americana*) to England from the Caribbean about 15 years after Buddle's death.

Classification

The genus *Buddleja* is now included in Scrophulariaceae, having earlier been classified under Buddlejaceae (synonym: Oftiaceae) and Loganiaceae of the approximately 100 species nearly all are shrubs <5 m (16 ft) tall, but a few qualify as trees, the largest reaching 30 m (98 ft). Both evergreen and deciduous species occur, in tropical and temperate regions resp. The leaves are lanceolate in most species, and arranged in opposite pairs on the stems (alternate in one species, *B. alternifolia*); they range from 1–30 cm (0.39–11.81 in) long. The flowers of the Asiatic species are mostly produced in terminal panicles 10–50 cm (3.9–19.7 in) long; the American species more commonly as cymes forming small, globose heads. Each individual flower is tubular and divided into four spreading lobes (petals) about 3–4 mm (0.12–0.16 in) across, the corolla length ranging from around 10 mm in the Asiatics to 3–30 mm in the American species, the wider variation in the latter because some South American species have evolved long red flowers to attract hummingbirds, rather than insects, as exclusive pollinators.

The colour of the flowers varies widely, from mostly pastel pinks and blues in Asia, to vibrant yellows and reds in the New World, while many cultivars have deeper tones. The flowers are generally rich in nectar and often strongly honey-scented. The fruit is a small capsule about 1 cm (0.39 in) long and 1–2 mm (0.039–0.079 in) diameter, containing numerous small seeds; in a few species (previously classified in the separate genus *Nicodemia*) the capsule is soft and fleshy, forming a berry.

Distribution

The genus is found in four continents. Over 60 species are native through the New World from the southern United States south to Chile, while many other species are found in the Old World, in Africa, and parts of Asia, but all are absent as natives from Europe and Australasia. The species are divided into three groups based on their floral type: Those in the New World are mostly dioecious (occasionally hermaphrodite or trioecious), while those in the Old World are exclusively hermaphrodite with perfect flowers.

Cultivation and Uses

As garden shrubs buddlejas are essentially 20th-century plants, with the exception of *B. globosa* which was introduced to Britain from southern Chile in 1774 and disseminated from the nursery of Lee and Kennedy, Hammersmith. Several species are popular garden plants, the species are commonly known as 'butterfly bushes' owing to their attractiveness to butterflies, and have become staples of the modern butterfly garden; they are also attractive to bees and moths.

The most popular cultivated species is *Buddleja davidii* from central China, named for the French Basque missionary and naturalist Père Armand David. Other common garden species include the aforementioned *B. globosa*, grown for its strongly honey-scented orange globular inflorescences, and the weeping *Buddleja alternifolia*. Several interspecific hybrids have been made, notably *B.* 'Lochinch' (*B. davidii* × *B. fallowiana*) and *B.* × *weyeriana* (*B. globosa* × *B. davidii*), the latter a cross between a South American and an Asiatic species.

Some species commonly escape from the garden. *B. davidii* in particular is a great coloniser of dry open ground; in urban areas in the United Kingdom, it often self-sows on waste ground or old masonry, where it grows into a dense thicket, and is listed as an invasive species in many areas. It is frequently seen beside railway lines, on derelict factory sites and, in the aftermath of World War II, on urban bomb sites. This earned it the popular nickname of 'the bombsite plant' among the war-time generation.

Popular garden cultivars include 'Royal Red' (reddish-purple flowers), 'Black Knight' (very dark purple), 'Sungold' (golden yellow), and 'Pink Delight' (pure pink). In recent years, much breeding work has been undertaken to create small, more compact buddlejas, such as 'Blue Chip' which reach no more than 2–3 ft (0.61–0.91 m) tall, and which are also seed sterile, an important consideration in the USA where *B. davidii* and its cultivars are banned from many states owing to their invasiveness.

CIRSIUM ARVENSE

Cirsium arvense is a perennial species of flowering plant in the family Asteraceae, native throughout Europe and northern Asia, and widely introduced elsewhere. The standard English name in its native area is creeping thistle. It is also commonly known as Canada thistle and field thistle.

The plant is beneficial for pollinators that rely on nectar. It also was a top producer of nectar sugar in a 2016 study in Britain, with a second-place ranking due to a production per floral unit of (2609 +/- 239 µg).

Alternative Names

A number of other names are used in other areas or have been used in the past, including: Canadian thistle, lettuce from hell thistle, California thistle, corn thistle, cursed thistle, field thistle, green thistle, hard thistle, perennial thistle, prickly thistle, small-flowered thistle, way thistle and stinger-needles. Canada and Canadian thistle are in wide use in the United States, despite being a misleading designation (it is not of Canadian origin).

Flowering creeping thistle.

Cirsium arvense is a C3 carbon fixation plant. The C3 plants originated during Mesozoic and Paleozoic eras, and tend to thrive in areas where sunlight intensity is moderate, temperatures are moderate, and ground water is plentiful. C_3 plants lose 97% of the water taken up through their roots to transpiration.

Creeping thistle is a herbaceous perennial plant growing up to 150 cm, forming extensive clonal colonies from thickened roots that send up numerous erect shoots during the growing season. It is a ruderal species.

Underground Network

Its underground structure consists of four types, 1) long, thick, horizontal roots, 2) long, thick, vertical roots, 3) short, fine shoots, and 4) vertical, underground stems. Though asserted in some literature, creeping thistle does not form rhizomes. Root buds form adventitiously on the thickened roots of creeping thistle, and give rise to new shoots. Shoots can also arise from the lateral buds on the underground portion of regular shoots, particularly if the shoots are cut off through mowing or when stem segments are buried.

Shoots and Leaves

Stems are 30–150 cm, slender green, and freely branched, smooth and glabrous (having no trichomes or glaucousness), mostly without spiny wings. Leaves are alternate on the stem with their base sessile and clasping or shortly decurrent. The leaves are very spiny, lobed, and up to 15–20 cm long and 2–3 cm broad (smaller on the upper part of the flower stem).

Flowers and Seeds

The inflorescence is 10–22 mm (0.39–0.87 in) in diameter, pink-purple, with all the florets of similar form (no division into disc and ray florets). The flowers are usually dioecious, but not invariably so, with some plants bearing hermaphrodite flowers. The seeds are 4–5 mm long, with a feathery pappus which assists in wind dispersal. One to 5 flower heads occur per branch, with plants in very favourable conditions producing up to 100 heads per shoot. Each head contains an average of 100 florets. Average seed production per plant has been estimated at 1530. More seeds are produced when male and female plants are closer together, as flowers are primarily insect-pollinated.

Pappus of *Cirsium arvense*.

A creeping thistle with a "cuckoo spit".

Varieties

Variation in leaf characters (texture, vestiture, segmentation, spininess) is the basis for determining creeping thistle varieties. According to Flora of Northwest Europe the two varieties are:

- *Cirsium arvense* var. *arvense*. Most of Europe. Leaves hairless or thinly hairy beneath.

- *Cirsium arvense* var. *incanum* (Fisch.) Ledeb. Southern Europe. Leaves thickly hairy beneath.

The Biology of Canadian Weeds: *Cirsium arvense* list four varieties:

- *Cirsium arvense* var. *vestitum* (Wimm. & Grab). Leaves gray-tomentose below.

- *Cirsium arvense* var. *integrifolium* (Wimm. & Grab). Leaves all entire or the upper leaves entire and the lower stem leaves shallowly and regularly pinnatifid or undulating.

- *Cirsium arvense* var. *arvense*. Leaves shallowly to deeply pinnatifid, often asymmetrical.

- *Cirsium arvense* var. *horridum* (Wimm. & Grab). Leaves thick, subcoriaceous, surface wavy, marginal spines long and stout.

Ecology

A European goldfinch (*Carduelis carduelis*) feeding on the seeds.

The seeds are an important food for the goldfinch and the linnet, and to a lesser extent for other finches. Creeping thistle foliage is used as a food by over 20 species of Lepidoptera, including the painted lady butterfly and the engrailed moth, and several species of aphids.

The flowers are visited by a wide variety of insects (the generalised pollination syndrome).

Status as a Weed

The species is widely considered a weed even where it is native, for example being designated an "injurious weed" in the United Kingdom under the Weeds Act 1959. It is also a serious invasive species in many additional regions where it has been introduced, usually accidentally as a contaminant in cereal crop seeds. It is cited as a noxious weed in several countries; for example Australia, Brazil, Canada, Ireland, New Zealand, and the United States. Many countries regulate this plant, or its parts (i.e., seed) as a contaminant of other imported products such as grains for consumption or seeds for propagation. In Canada, *C. arvense* is classified as a primary noxious weed seed in the Weed Seeds Order 2005 which applies to Canada's Seeds Regulations.

Control

Organic

Control methods include cutting at flower stem extension before the flower buds open to prevent seed spread. Repeated cutting at the same growth stage over several years may "wear down" the plant.

Growing forages such as alfalfa can help control the species as a weed by frequently cutting the alfalfa to add nutrients to the soil, the weeds also get cut, and have a harder time re-establishing themselves, which reduces the shoot density.

Orellia ruficauda feeds on Canada thistle and has been reported to be the most effective biological control agent for that plant. Its larvae parasitize the seed heads, feeding solely upon fertile seed heads.

The weevil *Larinus planus* also feeds on the thistle and has been used as a control agent in Canada. One larva of the species can consume up to 95% of seeds in a particular flower bud. However, use of this weevil has had a damaging effect on other thistle species as well, include some that are threatened. It may therefore not be a desirable control agent. It is unclear if the government continues to use this weevil to control Canada thistles or not.

The rust species *Puccinia obtegens* has shown some promise for controlling Canada thistle, but it must be used in conjunction with other control measures to be effective. Also *Puccinia punctiformis* is used in North America and New Zealand in biological control. In 2013, in four countries in three continents, epidemics of systemic disease caused by this rust fungus could be routinely and easily established. The procedure for establishing this control agent involves three simple steps and is a long-term sustainable control solution that is free and does not involve herbicides. Plants systemically diseased with the rust gradually but surely die. Reductions in thistle density were estimated, in 10 sites in the U.S., Greece, and Russia, to average 43%, 64%, and 81% by 18, 30, and 42 months, respectively, after a single application of spores of the fungus.

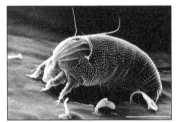
Electron scan micrography of Aceria anthocoptes.

Aceria anthocoptes feeds on this species and is considered to be a good potential biological control agent.

Chemical

Applying herbicide: Herbicides dominated by phenoxy compounds (especially MCPA) caused drastic declines in thistle infestation in Sweden in the 1950s. MCPA and clopyralid are approved in some regions.

Crop tolerance and weed control ratings were conducted in the spring of 2012, and the Prepass herbicide by DOW AgroSciences was found to be most effective at controlling the species as a weed problem in alfalfa fields.

Uses

Like other *Cirsium* species, the roots are edible, though rarely used, not in the least because of their propensity to induce flatulence in some people. The taproot is considered the most nutritious part. The leaves are also edible, though the spines make their preparation for food too tedious to be worthwhile. The stalks, however, are also edible and more easily despined. Bruichladdich distillery on Isle of Islay lists creeping thistle as one of the 22 botanical forages used in their gin, The Botanist.

The feathery pappus is also used by the Cherokee to fletch blowgun darts.

BRACHYPODIUM SYLVATICUM

Brachypodium sylvaticum is a perennial bunchgrass native to North Africa and Eurasia, and has recently been reported as rapidly invading coniferous forest understories in western Oregon. Occasionally cultivated for ornamental purposes, *B. sylvaticum* was first collected as an escaped invader in North America near Eugene, Oregon in 1939. By 1966, it was well-established in two large colonies near Corvallis, Oregon and since then has been quickly increasing in cover and range. It is now spreading into closed-canopy coniferous forests, riparian forests, forest edges, and upland prairies in full sun in Oregon's Willamette Valley and into the Cascade foothills. *Brachypodium sylvaticum* has the potential to spread throughout low elevation forests in the Pacific Northwest (Oregon, Washington, Idaho, British Columbia), and could spread into northern California too.

Brachypodium sylvaticum is a caespitose (tufted) perennial grass that typically grows 5 to 7 dm tall. It tends to form large clumps or bunches, but is apparently not rhizomatous. The hollow culms (stems) are pilose (i.e. bearing soft, spreading hairs) at the nodes and sometimes over the lower internodes. Its broad flat leaves are 4 to 10 mm wide, pilose, open-sheathed at the base, and do not have auricles. The ligules (1 to 2.5 mm long) are membranous, and are more or less erose-ciliolate.

Flowers of *B. sylvaticum* are located on pale green spikelets that are semi-spicate (i.e. the pedicels very short or lacking). Spikelets tend to be few (5 to 10), are 2 to 4 cm long, and are 7 to 17-flowered. The awns are straight, 10 to 15 mm long, and the unequal glumes are shorter than the florets that each one subtends. The lemmas are strongly ciliate and 2-nerved (2-veined). *Brachypodium sylvaticum* can be easily distinguished from *Bromus* species, in that the former has open leaf sheaths and spikelets which are either subsessile or short-pedicellate, while *Bromus* species have closed leaf sheaths and spikelets on long pedicels.

Range

Brachypodium sylvaticum is native to North Africa, northern and Mediterranean Europe, and Asia. In its native range, *B. sylvaticum* is most commonly found in forests and woodlands, but may also occur in open habitats.

As an invader in North America, *B. sylvaticum* is documented as invasive only in the state of Oregon. It occurs at sites ranging in elevation from sea level to about 1,200 m (4,000 ft) and it occupies a variety of aspects and light conditions. Within Oregon, *B. sylvaticum* is present in and around the Willamette Valley, and as far south as Josephine County (southwest Oregon near the California border). It currently covers thousands of acres in Oregon State University Research Forests near the city of Corvallis, and there are unconfirmed reports of this species in Colorado and Utah.

Analysis of the sites it already occupies in Oregon and its native range suggest *B. sylvaticum* has the potential to spread throughout low elevation forests in the Pacific Northwest (Oregon, Washington, Idaho, British Columbia), and could spread into northern California too.

Brachypodium sylvaticum is listed on the Pacific Northwest Exotic Pest Plant Council list B, indicating that it is a wildland weed of lesser invasiveness. This classification, however, may underestimate the threat it poses to native vegetation.

Reproduction

Brachypodium sylvaticum reproduces rapidly from seed, and although reportedly not rhizomatous, can resprout from small stem or root fragments when cut. It has also been suggested that *B. sylvaticum* does not maintain a persistent (longer than 1 year) seed bank in soils.

Management/Monitoring

Control methods for *B. sylvaticum* have not been well-studied. Removal of the entire plant by digging/hand removal is effective for small infestations, but is extremely time and labor-intensive. if enough of the root system is left in the soil, the plant will resprout.

Mowing or grazing treatments may control *B. sylvaticum*, if repeated for some time. In Europe, *B. sylvaticum* was absent in heavily grazed sites, indicating that repeated aboveground removal may eventually eliminate this species. Burning, however, seems ineffective, as *B. sylvaticum* is frequently found in recently burned sites . Repeated mowing, grazing, or burning treatments that are carried out before seed set may benefit control efforts by eliminating seed production each year, eventually exhausting the seed bank (if present). These methods may also increase the efficacy of subsequent herbicide treatments by forcing the plants to produce new shoots that are more likely to take up and be killed by herbicides.

Herbicide applications are currently the most effective technique known for controlling *B. sylvaticum*.

ALLIARIA PETIOLATA

Alliaria petiolata, or garlic mustard, is a biennial flowering plant in the mustard family (Brassicaceae). It is native to Europe, western and central Asia, north-western Africa, Morocco, Iberia and the British Isles, north to northern Scandinavia, and east to northern Pakistan and Xinjiang in western China.

In the first year of growth, plants form clumps of round, slightly wrinkled leaves, that when crushed smell like garlic. The next year plants flower in spring, producing cross shaped white flowers in dense clusters. As the flowering stems bloom they elongate into a spike-like shape. When blooming is complete, plants produce upright fruits that release seeds in mid-summer. Plants are often found growing along the margins of hedges, giving rise to the old British folk name of jack-by-the-hedge. Other common names include garlic mustard, garlic root, hedge garlic, sauce-alone, jack-in-the-bush, penny hedge and poor man's mustard. The genus name *Alliaria*, "resembling *Allium*", refers to the garlic-like odour of the crushed foliage.

All parts of the plant, including the roots, give off a strong odour like garlic.

It is an herbaceous biennial plant growing from a deeply growing, thin, whitish taproot scented like horseradish. In their first years, plants are rosettes of green leaves close to the ground; these rosettes remain green through the winter and develop into mature flowering plants the following spring. Second-year plants often grow from 30–100 cm (12–39 in) tall, rarely to 130 cm (51 in) tall. The leaves are stalked, triangular through heart shaped, 10–15 cm (3.9–5.9 in) long (of which about half being the petiole) and 5–9 cm (2.0–3.5 in) broad, with coarsely toothed margins. The flowers are produced in spring and summer in small clusters. Each small flower has four white petals 4–8 mm (0.2–0.3 in) long and 2–3 mm (0.08–0.12 in) broad, arranged in a cross shape. The fruit is an erect, slender, four-sided capsule 4–5.5 cm (1.6–2.2 in) long, called a silique, green maturing to pale grey brown, containing two rows of small shiny black seeds which are released when a silique splits open. A single plant can produce hundreds of seeds, which often scatter several meters from the parent plant.

Depending upon conditions, garlic mustard flowers either self-fertilize or are cross-pollinated by a variety of insects. Plants from self-fertilized seeds can be genetically identical to their parent plant, enhancing their abilities to thrive in places where their parental genotype can thrive.

Close-up of garlic mustard flowers.

Fruits and seeds.

Sixty-nine insect herbivores and seven fungi are associated with garlic mustard in Europe. The most important groups of natural enemies associated with garlic mustard were weevils (particularly the genus *Ceutorhynchus*), leaf beetles, butterflies, and moths, including the larvae of some moth species such as the garden carpet moth. The small white flowers have a rather unpleasant aroma which attracts midges and hoverflies, although the flowers usually pollinate themselves. In June the pale green caterpillar of the orange tip butterfly (*Anthocharis cardamines*) can be found feeding on the long green seed-pods from which it can hardly be distinguished.

Cultivation and Uses

Garlic mustard is one of the oldest spices used in Europe. Phytoliths in pottery of the Ertebølle and Funnelneck-Beaker culture in north-eastern Germany and Denmark, dating to 4100–3750 BCE prove its use. In the 17th century Britain, it was recommended as a flavouring for salt fish. It can also be made into a sauce for eating with roast lamb or salad. Early European settlers brought the herb to the New World to use as a garlic type flavouring. Its traditional medicinal purposes include use as a diuretic. The herb was also planted as a form of erosion control.

Today, the chopped leaves are used for flavouring in salads and sauces such as pesto, and sometimes the flowers and fruit are included as well. These are best when young, taste of both garlic and mustard. The seeds are sometimes used to season food in France. Garlic mustard was once used medicinally as a disinfectant or diuretic, and was sometimes used to treat wounds.

KNAPWEEDS

Knapweeds belong to the genus *Centaurea* and are members of the Sunflower family (Asteraceae). This is a very large and diverse family of plants that includes dandelions, sunflowers, and daisies. Most knapweeds are non-native to North America. They were brought to North America following the immigrant trail from Europe and Asia. Together, these Eurasian knapweeds form a large complex of invasive species that are found throughout the United States and Canada. All told, 25 species of knapweeds occur in the two countries, predominantly as noxious rangeland weeds in the West. Six species are considered in this manual. Among the most troublesome are diffuse, spotted, and squarrose knapweeds. Lesser-known knapweeds (meadow, brown, and black) are closely related to the others and are included in this manual because they share similar biology and some of the same biological control agents.

Knapweeds are highly invasive weeds that are capable of forming large infestations under favorable conditions. Knapweeds are distinguished by their bract shape, flower color, leaf shape, roots, seeds and branching habit.

Plant Development

All six knapweed species begin their lifecycle as seedlings that develop into prostrate rosettes of 5 to 12 lobed leaves. Most species remain rosettes the first year. With the onset of warm, moist conditions the following spring, plants bloom on one to several branched, flowering stems. Plants have many seedheads that occur singly at the tips of branched stems.

Like other members of the sunflower family, the knapweed head, or *capitulum*, is an aggregation of small, individual flowers. The individual flowers, or *florets*, are tightly clustered and anchored to a concave base, called the receptacle. The receptacle and florets are surrounded by an envelope of modified leaves, or *bracts*. Head size and bracts are important diagnostic characters for knapweeds.

As the head completes its development, the bracts separate to reveal the maturing florets, enabling pollination to occur. Seeds develop later in the season (knapweed seeds are also known as *achenes*). Seeds may have a tuft of whitish or tawny bristles at one end, called a *pappus*.

Insects used in knapweed biological control inflict damage to the plant in two places: The seedhead and the root. The plant is damaged by the larvae of these insects which feed in the head or root tissue, destroying it. Only the adult seedhead weevils eat foliage, otherwise adult insects generally don't damage the plant.

Seed-feeding biocontrol agents attack the plant at specific stages of development: Some attack the plant early, in the bud stage, and others attack later, when plants are in early to full bloom. Larvae eat and destroy seeds and receptacle tissue.

Root-boring biocontrol agents can attack the plant as soon as the root is large enough for the insect to feed. The root is composed of two key tissues: The root cortex and the central vascular tissue. Both tissues are nutritious: the cortex tissue stores nutrients and the vascular tissue contains the channels in which nutrients and water move up and down the plant.

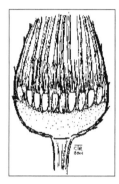 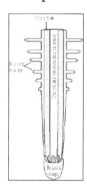

Knapweed Species

Bracts that surround the flower head are spine-tipped, biennial or short-lived perennial.

Central, terminal bract bent backwards (curved).

   Squarrose knapweed (*Centaurea virgata* ssp. *squarrosa*).

Central, terminal bract recurved.

   Diffuse knapweed (*Centaurea diffusa*).

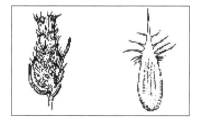

Flower heads without spine-tipped bracts.

Edge of bract is comb-like fringe.

Fringes of bracts short, drawn out and rigid, bract with brown triangular tip.

Spotted knapweed (*Centaurea stoebi*).

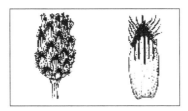

Fringes on bracts as long or longer than the width of the bract, not rigid.

Fringe on bract black.

Black knapweed (*Centaurea nigra*).

Fringe on bract tan to brown.

Meadow knapweed (*Centaurea pratensis*).

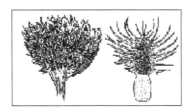

Bracts without comblike fringe, having a brown, papery, translucent tip.

Brown knapweed (*Centaurea jacea*).

Spotted Knapweed

Scientific name: *Centaurea stoebe* L. ssp. micranthos (Gugler) Hay/synonym C. biebersteinii L., formerly C. maculosa Lam.

A winter-hardy, short-lived perennial with deep taproots. Plants grow 6 to 24 inches (15 to 60 cm) in height and spread entirely by seeds. It is native to eastern Europe and Asia.

Leaves: The basal leaves are up to 8 inches (20 cm) long, deeply lobed, and arranged in a rosette. Stem leaves, arranged alternately, are smaller and not lobed. Uppermost leaves are bract-like.

Stems: The stems are upright, stiff, and branched. Small plants usually have an unbranched stem and one flower head; large plants have a stem with many branches and can have over 100 flower heads.

Flowers: Flowering occurs from June to October. The 0.2 to 0.4 inch (5 to 10 mm) long flower heads occur singly or in clusters at the branch tips. Each head bears stiff bracts, which are black-tipped, giving the plant its 'spotted' appearance. Heads contain from 30 to 50 pink or purple colored flowers.

Seeds: Seeds are 0.1 inch (2.5 mm) long, oval, black or brown with pale, vertical lines. Each seed has a short, bristly pappus about half the length of the seed. Plants can produce up to 600 seeds, some of which can remain dormant for many years.

Habitat and Occurrence: Spotted knapweed grows in a wide range of habitats, though mainly in grasslands and open forests. It has the widest distribution in the United States of all the knapweed species. A rapid colonizer of disturbed land, spotted knapweed can displace native vegetation in undisturbed areas. Heads persist on the stiff stems through the winter eventually breaking off when new rosette growth appears the following spring. Both diploid and tetraploid spotted knapweed types are known.

Diffuse Knapweed

Scientific name: *Centaurea diffusa* Lam.

A winter-hardy biennial or short-lived, tap-rooted perennial that reproduces entirely by seeds. Diffuse knapweed is originally from the eastern Mediterranean.

Leaves: The deeply lobed basal leaves are up to 4 inches (10 cm) long and 1 inch (2.5 cm) wide and arranged in a low-lying rosette. Lower stem leaves are alternate and divided into many lobes, whereas upper stem leaves are much smaller and have only a few slender lobes.

Stems: The single upright stem grows 6 to 24 inches (15 to 60 cm) in height and has numerous branches mostly on the upper half.

Flowers: Flowers are predominantly white, occasionally pink-purple. Heads are 0.5 inch (1.3 cm) long and covered with small, narrow bracts ending in sharp, rigid spines. The terminal spine is distinctly longer than the lateral, spreading spines. Flowering occurs from June to October.

Seeds: Seeds are 1/8 inch (5 mm) long, oblong, and dark brown. Seeds may have a pappus of short, pale bristles.

Habitat and Occurrence: Diffuse knapweed is wide-ranging, although it prefers habitats in the shrub-steppe zones and dry forest habitats. Though predominatly found in the Intermountain West, it is also found in the Midwest and the eastern U.S.

Like spotted knapweed, diffuse knapweed can displace native vegetation in undisturbed areas. Specialized chemicals give this weed a distinctive smell and an extremely bitter taste. Unlike other knapweeds, the heads of diffuse do not open to shed seeds. Instead, seeds are shed as the stiff, mature plants, tumble in the wind after the stiff central stalk breaks off. Seeds are also spread by vehicles, animals, and people. A diploid, fertile hybrid between diffuse knapweed and spotted knapweed has been identified. It is known as C. x psammogena.

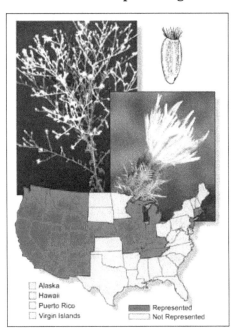

Squarrose Knapweed

Scientific name: *Centaurea virgata* Lam ssp. *squarrosa* Gigl.

Squarrose knapweed is a long-lived perennial with deep tap roots that reproduces only by seed. Squarrose knapweed came to the United States from the eastern Mediterranean.

Leaves: Rosettes of deeply lobed, gray-green leaves characterize squarrose knapweed.

Stems: The stems are upright, stiff, winged and branched. Small plants usually have an unbranched stem and one flower head; large plants have a stem with many branches and can have over 100 flower heads. Plants range in height from 6 to 24 inches (15 to 60 cm).

Flowers: Flowering occurs from July to September. Flower heads with 4 to 8 pink or purple flowers are borne singly or in pairs at the tips of branches. The seedheads are small and covered with spiny bracts having a long, recurved (backward pointing) terminal spine. The heads are deciduous, falling off the stems after the seeds mature.

Seeds: Squarrose knapweed seeds are pale to dark brown with pale vertical stripes and a short, white pappus. Only 3 to 4 seeds are produced per head, each measuring about 1/8 inch (5 mm) in length. Seeds are dispersed individually as they fall from the heads. Heads are transported when whole plants break off and tumble in the wind. Seeds disperse when whole heads break off from the stem and get lodged in the hair and fur of animals, much like cockleburs and burdock.

Habitat and Occurrence: Squarrose knapweed has a limited distribution in Utah, Oregon, California, Wyoming, and Michigan. It prefers dry, open rangeland with shallow soils.

Comments: Squarrose is similar to diffuse knapweed but has fewer flowers per head, recurved spines on the bracts, and is a true perennial.

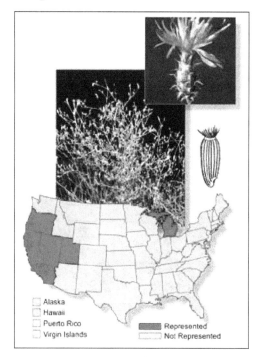

Meadow Knapweed

Scientific name: *Centaurea pratensis* Thuill.

Meadow knapweed is a deep-rooted perennial, growing each year from a woody root crown. It is native to Europe.

Leaves: Basal leaves are up to 6 inches (15.2 cm) long, tapering at both ends and having the broadest part above the middle of the leaf. Stem leaves are lance-shaped, shallowly-lobed and stalkless.

Stems: There are usually few to several stems with many branches. Mature plants reach 3.5 feet (1.04 m) tall.

Flowers: Flowers are generally rose-purple in color, although white flowers occasionally occur. Flowering occurs from July to September. The heads are solitary at the ends of the upper branches. They are broadly oval and almost globe-shaped, 0.5 inch (1.3 cm) long. The bracts of meadow knapweed are light to dark brown, with a fringed margin.

Seeds: Meadow knapweed seeds are pale tan in color, plumeless, 1/8 inch (2 cm) long.

Habitat and Occurrence: Meadow knapweed prefers moister and cooler conditions than the other knapweeds. It occurs predominantly in coastal Washington and Oregon, but is also found in moister, cooler habitats of the interior, e.g. forest openings along rivers and streams.

Comments: Meadow knapweed is a fertile hybrid between black and brown knapweeds.

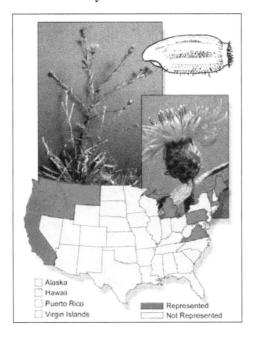

Black Knapweed

Scientific name: *Centaurea nigra* L.

Black knapweed is a perennial plant regrowing each year from a woody root crown. It was introduced into the United States from the United Kingdom.

Leaves: Basal rosette leaves are broad, stalked, and shallowly lobed. Stem leaves are smaller and not lobed.

Stems: Stems are erect and branched near the middle, from 8 to 32 inches (20 to 80 cm) tall, the base of the stem is sometimes prostrate and rooting from the nodes.

Flowers: Flowering occurs from July to October. Flowers are rose colored. Heads occur solitary at the ends of the upper branches. They are broad and rounded, 0.5 inch (1.3 cm) tall and 1 inch (2.54 cm) wide. The bracts of black knapweed are dark brown to black, with a comb-like fringe on the margin.

Seeds: Black knapweed produces about 60 seeds per head. They are ivory with lengthwise stripes, and have a pale, short pappus.

Habitat and Occurrence: Like meadow knapweed, black knapweed occurs predominantly in coastal Washington and Oregon, and in other cooler regions of the inland Northwest.

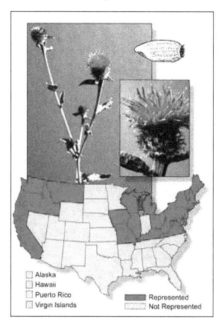

Brown Knapweed

Scientific name: *Centaurea jacea* L.

Brown knapweed is a perennial that reproduces only by seeds. It is native to Europe.

Leaves: Basal leaves are up to 6 inches (15.2 cm) long, tapering at both ends with the broadest part above the middle of the leaf. Stem leaves are lance-shaped, shallowly-lobed and stalkless.

Flowers: Flowers are rose-purple in color, rearely white. Flowering occurs from July to October. Heads are solitary at the ends of the upper branches. They are broadly oval. The bracts of brown knapweed are light to dark brown, with a papery, translucent margin.

Seeds: Brown knapweed seeds are light brown, plumeless, 1/8 inch (2 cm) long. Each head produces about twelve seeds.

Habitat and Occurrence: Like meadow knapweed, brown knapweed prefers moister, cooler conditions than the other knapweed species. It occurs predominantly in coastal Washington and Oregon, although it is distributed both in the West and the East. It also occurs in British Columbia.

KNOTWEEDS

Knotweeds *(Polygonum* spp.) are invasive perennials, with four species found in British Columbia: Japanese knotweed *(Fallopia japonica)*; Bohemian knotweed *(Fallopia* x *bohemica)*; Giant knotweed *(Fallopia sachalenensis)*; and Himalayan knotweed *(Polygonum polystachyum)*. Knotweeds thrive in roadside ditches, low-lying areas, irrigation canals, and other water drainage systems. They are also found in riparian areas, along stream banks, and in other areas with high soil moisture. Knotweeds occur in the southwest coastal region, the Shuswap, Kitimat, Stikine, Skeena, Columbia, Okanagan, and Kootenay areas, as well as the Queen Charlotte Islands. Additional plants may exist in many gardens in communities across BC.

Knotweeds have small white-green flowers that grow in showy, plume-like, branched clusters along the stem and leaf joints. Hollow stems stand upright and are bamboo-like with reddish-brown speckles and thin, papery sheaths. Leaves are heart or triangular-shaped on all species except Himalayan, which are elongated and tapered. Stems grow 1-5 metres in height at maturity, with leaves 8-10 centimetres wide and 15 centimetres in length. Giant knotweed leaves are generally twice the size of the other 3 species. A distinguishing feature of Japanese knotweed is the zigzag pattern in which leaves are arranged along the plant's arching stems.

Knotweeds spread rapidly through root systems that may extend from a parent plant up to 20 metres laterally and up to a depth of 3 metres. They thrive on freshly disturbed soil in moist locations. Knotweeds are dispersed by human activities or by water to downstream areas, and are of particular concern in riparian areas and areas prone to seasonal high water or flooding. Plants emerge in early spring and produce large leaves that can shade out other plant species. Infestations can dominate stream banks and reduce sight lines along roads, fences, and rights-of-way.

Knotweeds threaten biodiversity and disrupt the food chain by reducing available habitat and increasing soil erosion potential. Stream banks are at particular risk as exposed knotweed roots break off and float downstream to form new infestations. Knotweeds can reduce or eliminate access to water bodies for recreation activities including fishing, swimming, boating, canoeing, and kayaking.

A few native and ornamental alternatives to plant instead of knotweed include: Red-osier Dogwood; Black Elderberry; Peegee Hydrangea; False Soloman's Seal; and Goat's Beard.

RAGWEED

Ragweeds are flowering plants in the genus *Ambrosia* in the aster family, Asteraceae. They are distributed in the tropical and subtropical regions of the Americas, especially North America, where the origin and center of diversity of the genus are in the southwestern United States and northwestern Mexico. Several species have been introduced to the Old World and some have naturalized and have become invasive species. Ragweed species are expected to continue spreading across Europe in the near future in response to ongoing climate change.

Other common names include bursages and burrobrushes.

Ragweed pollen is notorious for causing allergic reactions in humans, specifically allergic rhinitis. Up to half of all cases of pollen-related allergic rhinitis in North America are caused by ragweeds.

The most widespread species of the genus in North America is *Ambrosia artemisiifolia*.

Ecology

Ragweeds are annual and perennial herbs and shrubs. Species may grow just a few centimeters tall or well exceed four meters in height. The stems are erect, decumbent or prostrate, and many grow from rhizomes. The leaves may be arranged alternately, oppositely, or both. The leaf blades come in many shapes, sometimes divided pinnately or palmately into lobes. The edges are smooth or toothed. Some are hairy, and most are glandular.

Ragweeds are monoecious, most producing inflorescences that contain both staminate and pistillate flowers. Inflorescences are often in the form of a spike or raceme made up mostly of staminate flowers with some pistillate clusters around the base. Staminate flower heads have stamens surrounded by whitish or purplish florets. Pistillate flower heads have fruit-yielding ovules surrounded by many phyllaries and fewer, smaller florets. The pistillate flowers are wind pollinated, and the fruits develop. They are burs, sometimes adorned with knobs, wings, or spines.

Many *Ambrosia* species occur in desert and semi-desert areas, and many are ruderal species that grow in disturbed habitat types.

Allergy

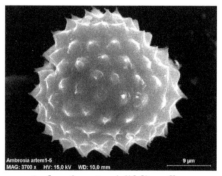

Ambrosia artemisiifolia pollen.

Ragweed pollen is a common allergen. A single plant may produce about a billion grains of pollen per season, and the pollen is transported on the wind. It causes about half of all cases of pollen-associated allergic rhinitis in North America, where ragweeds are most abundant and diverse. Common culprits are common ragweed (*A. artemisiifolia*) and great ragweed (*A. trifida*).

Concentration of ragweed pollen—in the absence of significant rainfall, which removes pollen from the air—is the lowest in the early morning hours (6:00 AM), when emissions starts, and pollen concentration peaks at midday. Ragweed pollen can remain airborne for days and travel great distances, affecting people hundreds of miles away. It can even be carried 300 to 400 miles (640 km) out to sea. Ragweeds native to the Americas have been introduced to Europe starting in the nineteenth century and especially during World War I, and have spread rapidly there since the 1950s. Eastern Europe, particularly Hungary, has been badly affected by ragweed since the early 1990s, when the dismantling of Communist collective agriculture led to large-scale abandonment of agricultural land, and new building projects also resulted in disturbed, un-landscaped acreage.

The major allergenic compound in the pollen has been identified as *Amb a 1*, a 38 kDa nonglycosylated protein composed of two subunits. It also contains other allergenic components, such as profilin and calcium-binding proteins.

Ragweed allergy sufferers may show signs of oral allergy syndrome, a food allergy classified by a cluster of allergic reactions in the mouth in response to the consumption certain fruits, vegetables, and nuts. Foods commonly involved include beans, celery, cumin, hazelnuts, kiwifruit, parsley, potatoes, bananas, melons, cucumbers, and zucchini. Because cooking usually denatures the proteins that cause the reaction, the foods are more allergenic when eaten raw; exceptions are celery and nuts, which may not be safe even when cooked. Signs of reaction can include itching, burning, and swelling of the mouth and throat, runny eyes and nose, hives, and, less commonly, vomiting, diarrhea, asthma, and anaphylaxis. These symptoms are due to the abnormal increase of IgE antibodies which attach to a type of immune cell called mast cells. When the ragweed antigen then attaches to these antibodies the mast cells release histamine and other symptom evoking chemicals.

Merck & Co, under license from allergy immunotherapy (AIT) company ALK, has launched a ragweed allergy immunotherapy treatment in sublingual tablet form in the US and Canada. Allergy immunotherapy treatment involves administering doses of the allergen to accustom the body to induce specific long-term tolerance.

Control and Eradication

Chemical spraying has been used for control in large areas. Because ragweed only reacts to some of the more aggressive herbicides, it is highly recommended to consult professionals when deciding on dosage and methodology, especially near urban areas. Effective active ingredients include those that are glyphosate-based (Roundup, Glyphogan, Glialka), sulfosate-based (Medallon), and glufosinate ammonium-based (Finale 14SL). In badly infested areas, 2 to 6.5 liters per hectare (0.2–0.7 U.S. gal/acre) are usually dispersed. In 2007 several *Ambrosia artemisiifolia* populations were glyphosate resistant, exclusively in the USA.

Where herbicides cannot be used, mowing may be repeated about every three weeks, as it grows back rapidly. In the past, ragweed was usually cut down, left to dry, and then burned. This method is used less often now, because of the pollution caused by smoke. Manually uprooting ragweed is generally ineffective, and skin contact can cause allergic reaction. If uprooting is the method of choice, it should be performed before flowering. There is evidence that mechanical and chemical control methods are actually no more effective in the long run than leaving the weed in place.

Fungal rusts and the leaf-eating beetle *Ophraella communa* have been proposed as agents of biological pest control of ragweeds, but the latter may also attack sunflowers, and applications for permits and funding to test these controls have been unsuccessful. The beetle has, however, appeared in Europe, either on its own or as an uncontrolled introduction, and it has started making a dent into Ambrosia populations there.

Species

Ambrosia dumosa.

Ambrosia chamissonis.

Ambrosia ambrosioides.

There are about 50 species in genus *Ambrosia*. Species include:

- *Ambrosia acanthicarpa*
- *Ambrosia acuminata*
- *Ambrosia ambrosioides*
- *Ambrosia arborescens*
- *Ambrosia artemisiifolia*
- *Ambrosia artemisioides*
- *Ambrosia bidentata*
- *Ambrosia bryantii*
- *Ambrosia camphorata*
- *Ambrosia canescens*
- *Ambrosia carduacea*
- *Ambrosia chamissonis*
- *Ambrosia cheiranthifolia*
- *Ambrosia chenopodiifolia*
- *Ambrosia confertiflora*
- *Ambrosia cordifolia*
- *Ambrosia deltoidea*

- *Ambrosia dentata*
- *Ambrosia divaricata*
- *Ambrosia diversifolia*
- *Ambrosia dumosa*
- *Ambrosia eriocentra*
- *Ambrosia flexuosa*
- *Ambrosia grayi*
- *Ambrosia × helenae*
- *Ambrosia hispida*
- *Ambrosia humi*
- *Ambrosia ilicifolia*
- *Ambrosia × intergradiens*
- *Ambrosia johnstoniorum*
- *Ambrosia linearis*
- *Ambrosia magdalenae*
- *Ambrosia maritima*
- *Ambrosia microcephala*

- *Ambrosia monogyra*
- *Ambrosia nivea*
- *Ambrosia pannosa*
- *Ambrosia peruviana*
- *Ambrosia × platyspina*
- *Ambrosia polystachya*
- *Ambrosia psilostachya*
- *Ambrosia pumila*
- *Ambrosia salsola*
- *Ambrosia scabra*
- *Ambrosia tacorensis*
- *Ambrosia tarapacana*
- *Ambrosia tenuifolia*
- *Ambrosia tomentosa*
- *Ambrosia trifida*
- *Ambrosia velutina*
- *Ambrosia villosissima*

ENGLISH IVY

Hedera helix, the common ivy, English ivy, European ivy, or just ivy, is a species of flowering plant in the family Araliaceae, native to most of Europe and western Asia. A rampant, clinging evergreen vine, it is a familiar sight in gardens, waste spaces, on walls, tree trunks and in wild areas across its native habitat.

Ivy growing on a granite crag.

Stems showing the rootlets used to cling to walls and tree trunks.

Hedera helix is an evergreen climbing plant, growing to 20–30 m (66–98 ft) high where suitable surfaces (trees, cliffs, walls) are available, and also growing as groundcover where no vertical surfaces occur. It climbs by means of aerial rootlets with matted pads which cling strongly to the substrate. The ability to climb on surfaces varies with the plants variety and other factors, *Hedera helix* prefers non-reflective, darker and rough surfaces with near-neutral pH. It generally thrives in a wide range of soil pH with 6.5 being ideal, prefers moist, shady locations and avoids exposure to direct sunlight, the latter promoting drying out in winter.

The leaves are alternate, 50–100 mm (2–4 in) long, with a 15–20 mm (0.6–0.8 in) petiole; they are of two types, with palmately five-lobed juvenile leaves on creeping and climbing stems, and un-lobed cordate adult leaves on fertile flowering stems exposed to full sun, usually high in the crowns of trees or the top of rock faces.

The flowers are produced from late summer until late autumn, individually small, in 3-to-5 cm-diameter (1.2-to-2.0 in) umbels, greenish-yellow, and very rich in nectar, an important late autumn food source for bees and other insects.

The fruit are purple-black to orange-yellow berries 6–8 mm (0.2–0.3 in) in diameter, ripening in late winter, and are an important food for many birds, though somewhat poisonous to humans.

One to five seeds are in each berry, which are dispersed after being eaten by birds.

Subspecies

The three subspecies are:

- *H. h. helix*: Central, northern and western Europe, plants without rhizomes, purple-black ripe fruit.

- *H. h. poetarum* Nyman (syn. *Hedera chrysocarpa* Walsh): Southeast Europe and southwest Asia (Italy, Balkans, Turkey), plants without rhizomes, orange-yellow ripe fruit.

- *H. h. rhizomatifera* McAllister: Southeast Spain, plants rhizomatous, purple-black ripe fruit.

The closely related species *Hedera canariensis* and *Hedera hibernica* are also often treated as subspecies of *H. helix*, though they differ in chromosome number so do not hybridise readily. *H. helix* can be best distinguished by the shape and colour of its leaf trichomes, usually smaller and slightly more deeply lobed leaves and somewhat less vigorous growth, though identification is often not easy.

Range

Common ivy clinging on a plane tree.

It ranges from Ireland northeast to southern Scandinavia, south to Portugal, and east to Ukraine and Iran and northern Turkey.

The northern and eastern limits are at about the −2 °C (28 °F) winter isotherm, while to the west and southwest, it is replaced by other species of ivy. *Hedera helix* itself is much more winter-hardy and survives temperatures of −23.3 °C (−9.9 °F) and above.

Cultivation and Uses

Ivy-covered entrance.

Variegated leaves at Enchanting Floral Gardens.

Ivy is widely cultivated as an ornamental plant. Within its native range, the species is greatly valued for attracting wildlife. The flowers are visited by over 70 species of nectar-feeding insects, and the berries eaten by at least 16 species of birds. The foliage provides dense evergreen shelter, and is also browsed by deer.

In Europe, it is frequently planted to cover walls and the government recommends growing it on buildings for its ability to cool the interior in summer, while providing insulation in winter, as well as protecting the covered building from soil moisture, temperature fluctuations and direct exposure to heavy weather. Further uses include weed suppression in plantings, beautifying unsightly facades and providing additional green by growing on tree trunks.

However, ivy can be problematic. It is a fast-growing, self-clinging climber that is capable of causing damage to brickwork, guttering, etc., and hiding potentially serious structural faults, as well as harbouring unwelcome pests. Careful planning and placement are essential.

Cultivars

Over 30 cultivars have been selected for such traits as yellow, white, variegated (e.g., 'Glacier'), and deeply lobed leaves (e.g. 'Sagittifolia'), purple stems, and slow, dwarfed growth.

The following cultivars have gained the Royal Horticultural Society's Award of Garden Merit:

- 'Angularis Aurea'
- 'Buttercup'
- 'Caecilia'
- 'Ceridwen'
- 'Congesta'
- 'Duckfoot'
- 'Glacier'
- 'Goldchild'
- 'Golden Ingot'
- 'Manda's Crested'
- 'Midas Touch'
- 'Parsley Crested'
- 'Shamrock'
- 'Spetchley'

Ethnomedical Uses

Ivy extracts are part of current cough medicines. In the past, the leaves and berries were taken orally as an expectorant to treat cough and bronchitis. In 1597, the British herbalist John Gerard recommended water infused with ivy leaves as a wash for sore or watering eyes. The leaves can cause severe contact dermatitis in some people. People who have this allergy (strictly a type IV hypersensitivity) are also likely to react to carrots and other members of the Apiaceae as they contain the same allergen, falcarinol.

Invasive Species

Like other exotic species, ivy has predominantly been spread to areas by human action. *H. helix* is labeled as an invasive species in many parts of the United States, and its sale or import is banned in the state of Oregon.

Having disappeared during the glaciation, ivy is believed to have been spread back across the continent by birds once the continent warmed up again. With a great capacity for adaptation, ivy will grow wherever development conditions and habitat similar to that of its European origins exist, occurring as opportunistic species across a wide distribution with close vicariant relatives and few species, indicating recent speciation.

Australia

Hedera used decoratively as underplanting.

It is considered a noxious weed across southern, especially south-eastern, Australia and local councils provide free information and limited services for removal. In some councils it is illegal to sell the plant. It is a weed in the Australian state of Victoria.

New Zealand

H. helix has been listed as an "environmental weed" by the Department of Conservation since 1990.

United States

In the United States, *H. helix* is considered weedy or invasive in a number of regions and is on the official noxious weed lists in Oregon and Washington. Like other invasive vines such as kudzu, *H. helix* can grow to choke out other plants and create "ivy deserts". State- and county-sponsored efforts are encouraging the destruction of ivy in forests of the Pacific Northwest and the Southern United States. Its sale or import is banned in Oregon. Ivy can easily escape from cultivated gardens and invade nearby parks, forests and other natural areas.

Control and Eradication

Ivy should not be planted or encouraged in areas where it is invasive. Where it is established, it is very difficult to control or eradicate. In the absence of active and ongoing measures to control its growth, it tends to crowd out all other plants, including shrubs and trees.

Damage to Trees

Ivy can climb into the canopy of young or small trees in such density that the trees fall over from the weight, a problem that does not normally occur in its native range. In its mature form, dense ivy can destroy habitat for native wildlife and creates large sections of solid ivy where no other plants can develop.

Use as Building Facade Green

As with any self-climbing facade green, some care is required to make best use of the positive effects: Ivy covering the walls of an old building is a familiar and often attractive sight. It has insulating as well as weather protection benefits, dries the soil and prevents wet walls, but can be problematic if not managed correctly.

Ivy, and especially European ivy (*H. helix*) grows vigorously and clings by means of fibrous roots, which develop along the entire length of the stems. These are difficult to remove, leaving an unsightly "footprint" on walls, and possibly resulting in expensive resurfacing work. Additionally, ivy can quickly invade gutters and roofspaces, lifting tiles and causing blockages. It also harbors mice and other creatures. The plants have to be cut off at the base, and the stumps dug out or killed to prevent regrowth.

Therefore, if a green facade is desired, this decision has to be made consciously, since later removal would be tedious.

Mechanism of Attachment

Hedera helix is able to climb relatively smooth vertical surfaces, creating a strong, long lasting adhesion with a force of around 300 nN. This is accomplished through a complex method of attachment starting as adventitious roots growing along the stem make contact with the surface and extend root hairs that range from 20-400 μm in length. These tiny hairs grow into any small crevices available, secrete glue-like nanoparticles, and lignify. As they dry out, the hairs shrink and curl, effectively pulling the root closer to the surface. The glue-like substance is a nano composite adhesive that consists of uniform spherical nanoparticles 50-80 nm in diameter in a liquid polymer matrix. Chemical analyses of the nanoparticles detected only trace amounts of metals, once thought to be responsible for their high strength, indicating that they are largely organic. Recent work has shown that the nanoparticles are likely composed in large part of arabinogalactan proteins (AGPs), which exist in other plant adhesives as well. The matrix portion of the composite is made of pectic polysaccharides. Calcium ions present in the matrix induce interactions between carboxyl groups of these components, causing a cross linking that hardens the adhesive.

MIMOSA PUDICA

Mimosa pudica also called sensitive plant, sleepy plant, action plant, Dormilones, touch-me-not, shameplant, zombie plant, or shy plant) is a creeping annual or perennial flowering plant of the pea/legume family Fabaceae and Magnoliopsida taxon, often grown for its curiosity value: the compound leaves fold inward and droop when touched or shaken, defending themselves from harm, and re-open a few minutes later. In the UK it has gained the Royal Horticultural Society's Award of Garden Merit.

The species is native to South America and Central America, but is now a pantropical weed, and can be found in Southern United States, South Asia, East Asia and South Africa as well. It is not shade tolerant, and is primarily found on soils with low nutrient concentrations *Mimosa pudica* is well known for its rapid plant movement. Like a number of other plant species, it undergoes changes in leaf orientation termed "sleep" or nyctinastic movement. The foliage closes during darkness and reopens in light. This was first studied by the French scientist Jean-Jacques d'Ortous de Mairan. Due to *Mimosa's* unique response to touch, it became an ideal plant for many experiments regarding plant habituation and memory.

Mimosa pudica flower.

Flower.

Mimosa pudica folding leaflets inward.

Mimosa pudica seeds.

Mimosa pudica with mature seed pods on plant.

Mimosa pudica seedling with two cotyledons and some leaflets.

The whole plant of *Mimosa pudica* includes thorny stem and branches, flower head, dry flowers, seed pods, and folded and unfolded leaflets.

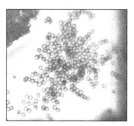

Pollens.

The stem is erect in young plants, but becomes creeping or trailing with age. It can hang very low and become floppy. The stem is slender, branching, and sparsely to densely prickly, growing to a length of 1.5 m (5 ft).

The leaves are bipinnately compound, with one or two pinnae pairs, and 10–26 leaflets per pinna. The petioles are also prickly. Pedunculate (stalked) pale pink or purple flower heads arise from the leaf axils in mid summer with more and more flowers as the plant gets older. The globose to ovoid heads are 8–10 mm (0.3–0.4 in) in diameter (excluding the stamens). On close examination, it is seen that the floret petals are red in their upper part and the filaments are pink to lavender.

The fruit consists of clusters of two to eight pods from 1–2 cm (0.4–0.8 in) long each, these being prickly on the margins. The pods break into two to five segments and contain pale brown seeds about 2.5 mm (0.1 in) long. The flowers are insect pollinated and wind pollinated. The seeds have hard seed coats which restrict germination and make osmotic pressure and soil acidity less significant hindrances. High temperatures are the main stimuli that cause the seeds to end dormancy.

The roots of *Mimosa pudica* create carbon disulfide, which prevents certain pathogenic and mycorrhizal fungi from growing within the plant's rhizosphere. This allows the formation of nodules on the roots of the plant that contain endosymbiotic diazotrophs, which fix atmospheric nitrogen and convert it into a form that is usable by the plant.

Mimosa pudica is a tetraploid ($2n = 52$).

Plant movement

The leaflets also close when stimulated in other ways, such as touching, warming, blowing, shaking, which are all encapsulated within mechanical or electrical stimulation. These types of movements have been termed seismonastic movements. This reflex may have evolved as a defense mechanism to disincentivize predators, or alternatively to shade the plant in order to reduce water lossage due to evaporation. The main structure mechanistically responsible for the drooping of the leaves is the pulvinus. The stimulus is transmitted as an action potential from a stimulated leaflet, to the leaflet's swollen base (pulvinus), and from there to the pulvini of the other leaflets, which run along the length of the leaf's rachis. The action potential then passes into the petiole, and finally to the large pulvinus at the end of the petiole, where the leaf attaches to the stem. The pulvini cells gain and lose turgor due to water moving in and out of these cells, and multiple ion concentrations play a role in the manipulation of water movement.

Ions cannot easily move in and out of cells, so protein channels such as voltage-gated potassium channels and calcium-permeable anion channels are responsible for allowing potassium and calcium, respectively, to flow through the cell membrane, making cells permeable to these ions. The action potential causes potassium ions to flow out from the vacuoles of cells in the various pulvini. Differences in turgidity in different regions of the leaf and stem results in the closing of the leaflets and the collapse of the leaf petiole. Other important proteins include H+-ATPases, aquaporins, and actin, which all aid in the redistribution of ions in the pulvini, especially during a seismonastic response. H+-ATPases and aquaporins aid in the direct movement of water molecules, while actin's role has a more biochemical explanation. Actin is composed of many phosphorylated tyrosine (an amino acid) molecules, and manipulation of how phosphorylated the tyrosine molecules are directly correlates to how much the *M. pudica* leaves droop.

This movement of folding inwards is energetically costly for the plant and also interferes with the process of photosynthesis. This characteristic is quite common within the Mimosoideae subfamily of the legume family, Fabaceae. The stimulus can also be transmitted to neighboring leaves. It is not known exactly why *Mimosa pudica* evolved this trait, but many scientists think that the plant uses its ability to shrink as a defense from herbivores. Animals may be afraid of a fast moving plant and would rather eat a less active one. Another possible explanation is that the sudden movement dislodges harmful insects.

The movement of calcium, potassium, and chloride ions in pulvini cells has been analyzed to better understand how ion and water flux affect *M. pudica* leaves drooping. A batch of *M. pudica* were grown and watered daily, and 10-20 pulvini were collected from each group of pulvini reactive to touch, and pulvini unreactive to touch. To further understand the movement of the ions, the upper and lower halves of all collected pulvini underwent separate ion analysis using the x-ray fluorescence spectroscopy method. This method tracked the location of the ions by coloring them each with a different color of fluorescence dye. In terms of calcium concentrations, there was a significant difference in concentration between both halves of the pulvini, in both the reactive and nonreactive pulvini. For potassium, higher concentrations were found throughout the reactive pulvini, while a high concentration difference in the top and bottom halves of the unreactive pulvini was found.

Upon stimulation to the reactive pulvini, an average of a 240% increase in potassium was found in the pulvini cells. Analysis of chloride ions gave similar results where high concentrations were found throughout the reactive pulvini, except in the unreactive pulvini the difference in concentration in the top and bottom pulvini was not significant. Overall, what was found was that high concentrations of potassium, chloride, and calcium led to a rapid decrease in water in the pulvini, which results in the drooping of the *M. pudica* leaves. Mitigation of this gradient resulted in less reaction and movement of the leaves.

Distribution and Habitat

Mimosa pudica is native to South America, North America and Central America. It can also be found in Asia in countries such as Singapore, Bangladesh, Thailand, India, Nepal, Indonesia, Malaysia, Philippines, Vietnam, Cambodia, Laos, Japan, Sri Lanka, Guam and well across the southern part of the United States. It has been introduced to many other regions and is regarded as an invasive species in Tanzania, South Asia and South East Asia and many Pacific Islands. It is regarded as invasive in parts of Australia and is a declared weed in the Northern Territory, and Western Australia although not naturalized there. Control is recommended in Queensland.

It has also been introduced to Uganda, Ghana, Nigeria, Seychelles, Mauritius and East Asia but is not regarded as invasive in those places. In the United States of America, it grows in Louisiana, Florida, Hawaii, Tennessee, Virginia, Maryland, Puerto Rico, Texas, Alabama, Mississippi, North Carolina, Georgia, and the Virgin Islands, as well as Cuba and the Dominican Republic.

Predators

Mimosa pudica has several natural predators, such as the Spider mite and Mimosa webworm. Both of these insects wrap the leaflets in webs that hinder the responsive closing. Webbed leaves are noticeable as they become brown fossilized remnants after an attack. The Mimosa webworm is composed of two generations that arise at different seasons. This makes prevention difficult and requires proper timing of insecticides to avoid aiding other predators. Once the larvae become steel-gray moths they are harmless to the plant, but lay more eggs.

Agricultural Impacts

The species can be a weed for tropical crops, particularly when fields are hand cultivated. Crops it tends to affect are corn, coconuts, tomatoes, cotton, coffee, bananas, soybeans, papaya, and sugar cane. Dry thickets may become a fire hazard. In some cases it has become a forage plant although the variety in Hawaii is reported to be toxic to livestock.

In addition, *Mimosa pudica* can change the physico-chemical properties of the soil it invades. For example, the total N and K increased in significantly invaded areas.

Phytoremediation

36 native Thai plant species were tested to see which conducted the most phytoremediation of arsenic polluted soils caused by tin mines. *Mimosa pudica* was one of the four species that significantly extracted and bioaccumulated the pollutant into its leaves. Other studies have found that *Mimosa pudica* extracts heavy metals such as copper, lead, tin, and zinc from polluted soils. This allows for the soil to gradually return to less toxic compositions.

Nitrogen Fixation

Mimosa pudica can form root nodules that are habitable by nitrogen-fixing bacteria. The bacteria are able to convert atmospheric nitrogen, which plants cannot use, into a form that plants can use. This trait is common among plants in the family Fabaceae. Nitrogen is a vital element for both plant growth and reproduction. Nitrogen is also essential for plant photosynthesis because it is a component of chlorophyll. Nitrogen fixation contributes nitrogen to the plant and to the soil surrounding the plant's roots.

Mimosa pudica's ability to fix nitrogen may have arisen in conjunction with the evolution of nitrogen-fixing bacteria. Nitrogen fixation is an adaptive trait that has transformed the parasitic relationship between the bacteria and plants into a mutualistic relationship. The shifting dynamics of this relationship are demonstrated by the corresponding improvement of various symbiotic charactcristics in both *Mimosa pudica* and bacteria. These traits include enhanced "competitive nodulation, nodule development, intracellular infection, and bacteroid persistence".

As much as 60% of the nitrogen found in *Mimosa pudica* can be attributed to the fixation of N_2 by bacteria. *Burkholderia phymatum* STM815^T and *Cupriavidus taiwanensis* LMG19424^T are beta-rhizobial strains of diazotrophs that are highly effective at fixing nitrogen when coupled with *M. pudica*. *Burkholderia* is also shown to be a strong symbiont of *Mimosa pudica* in nitrogen-poor soils in regions like Cerrado and Caatinga.

Cultivation

In cultivation, this plant is most often grown as an indoor annual, but is also grown for groundcover. Propagation is generally by seed. *Mimosa pudica* grows most effectively in nutrient poor soil that allows for substantial water drainage. However, this plant is also shown to grow in scalped and eroded subsoils. Typically, disrupted soil is necessary in order for *M. pudica* to become established in an area. Additionally, the plant is shade intolerant and frost-sensitive, meaning that it does not tolerate low levels of light or cold temperatures. *Mimosa pudica* does not compete for resources with larger foliage or forest canopy undergrowth.

In temperate zones it must be grown under protection, where the temperature does not fall below 13 °C (55 °F).

Chemical Constituents

Mimosa pudica contains the toxic alkaloid mimosine, which has been found to also have antiproliferative and apoptotic effects. The extracts of *Mimosa pudica* immobilize the filariform larvae of *Strongyloides stercoralis* in less than one hour. Aqueous extracts of the roots of the plant have shown significant neutralizing effects in the lethality of the venom of the monocled cobra (*Naja kaouthia*). It appears to inhibit the myotoxicity and enzyme activity of cobra venom.

Mimosa pudica demonstrates both antioxidant and antibacterial properties. This plant has also been demonstrated to be non-toxic in brine shrimp lethality tests, which suggests that *M. pudica* has low levels of toxicity. Chemical analysis has shown that *Mimosa pudica* contains various compounds, including "alkaloids, flavonoid C-glycosides, sterols, terenoids, tannins, and fatty acids". The roots of the plant have been shown to contain up to 10% tannin. A substance similar to adrenaline has been found within the plant's leaves. *Mimosa pudica*'s seeds produce mucilage made up of D-glucuronic acid and D-xylose. Additionally, extracts of *M. pudica* have been shown to contain crocetin-dimethylester, tubulin, and green-yellow fatty oils. A new class of phytohormone turgorines, which are derivatives of gallic acid 4-O-(β-D-glucopyranosyl-6'-sulfate), have been discovered within the plant.

The nitrogen-fixing properties of *Mimosa pudica* contribute to a high nitrogen content within the plant's leaves. The leaves of *M. pudica* also contain a wide range of carbon to mineral content, as well as a large variation in ^{13}C values. The correlation between these two numbers suggests that significant ecological adaptation has occurred among the varieties of *M. pudica* in Brazil.

The roots contain sac-like structures that release organic and organosulfur compounds including SO_2, methylsulfinic acid, pyruvic acid, lactic acid, ethanesulfinic acid, propane sulfinic acid, 2-mercaptoaniline, S-propyl propane 1-thiosulfinate, and thioformaldehyde, an elusive and highly unstable compound never before reported to be emitted by a plant.

EUPHORBIA LATHYRIS

Euphorbia lathyris, the caper spurge or paper spurge, is a species of spurge native to southern Europe, France, Italy, Greece, northwest Africa, and eastward through southwest Asia to western China.

Other names occasionally used include gopher spurge, gopher plant or mole plant.

Growth

It is an erect biennial (occasionally annual) plant growing up to 1.5 m tall, with a glaucous blue-green stem. The leaves are arranged in decussate opposite pairs, and are lanceolate, 5–15 cm long and 1-2.5 cm broad, glaucous blue-green with a waxy texture and pale greenish-white midrib and veins. The flowers are green to yellow-green, 4 mm diameter, with no petals. The seeds are green ripening brown or grey, produced in globular clusters 13–17 mm diameter of three seeds compressed together.

Chemical Characteristics

All parts of the plant, including the seeds and roots are poisonous. Handling may cause skin irritation as the plant produces latex. While poisonous to humans and most livestock, goats sometimes eat it and are immune to the toxin. However, the toxin can be passed through the goat's milk.

Habitat

Away from its native range, it is widely naturalised in many regions, where it is often considered an invasive weed. It grows in partial shade to full sun in USDA zones 5–9.

Uses

The mole plant is sold by some nurseries as it is believed to repel moles.

It is used in folk medicine as a remedy for cancer, corns, and warts and has been used by beggars to induce skin boils.

ALLIUM VINEALE

Allium vineale (wild garlic, onion grass, crow garlic or stag's garlic) is a perennial, bulb-forming species of wild onion, native to Europe, northwestern Africa and the Middle East. The species was introduced in Australia and North America, where it has become a noxious weed.

All parts of the plant have a strong garlic odour. The underground bulb is 1–2 cm diameter, with a fibrous outer layer. The main stem grows to 30–120 cm tall, bearing 2–4 leaves and an apical inflorescence 2–5 cm diameter comprising a number of small bulbils and none to a few flowers, subtended by a basal bract. The leaves are slender hollow tubes, 15–60 cm long and 2–4 mm thick, waxy texture, with a groove along the side of the leaf facing the stem. The inflorescence is a tight umbel surrounded by a membranous bract in bud which withers when the flowers open. Each individual flower is stalked and has a pinkish-green perianth 2.5 to 4.5 mm (0.10 to 0.18 in) long. There are six tepals, six stamens and a pistil formed from three fused carpels. Mixed with the flowers are several of yellowish-brown bulbils. The fruit is a capsule but the seeds seldom set and propagation usually takes place when the bulbils are knocked off and grow into new plants. Plants with no flowers, only bulbils, are sometimes distinguished as the variety *Allium vineale* var. *compactum*, but this character is probably not taxonomically significant.

Uses and Problems

While *Allium vineale* has been suggested as a substitute for garlic, there is some difference of opinion as to whether there is an unpleasant aftertaste compared to that of common garlic (*Allium sativum*). It imparts a garlic-like flavour and odour on dairy and beef products when grazed by livestock. It is considered a pestilential invasive weed in the USA, as grain products may become tainted with a garlic odour or flavour in the presence of aerial bulblets at the time of harvest. Wild garlic is tolerant to herbicides, which cannot cling well to the vertical, smooth and waxy structure of its leaves.

EQUISETUM

Equisetum is a genus of ferns commonly known as the 'horsetails'. They consist of 15 species of considerably unique plants from the class of ferns known as Equisetopsida.

Horsetails have a distinct growth form and spores with the ability to travel along the ground. They are found through most parts of Earth in wet environments.

Growth Form

Horsetails have a unique growth form that distinguishes them amongst all other ferns. Their stems, leaves and roots are all quite unique and makes the *Equisetum* species a very interesting group of plants.

Their stems often appear to be formed by the combination of multiple smaller segments and they often resemble the stems of rushes. Stems grow either as straight erect stalks or creeping along the ground; they are hollow and reinforced with silica. The silica makes the stems of horsetails strong and hardened.

Equisetum leaves grow in whorls with multiple leaves growing from the same point around the stem and their branches also grow in the same fashion. Some species have tiny scale-like leaves and often appear leafless, whilst other species have long, slender leaves and give the plant a feathery appearance. The particular points where leaves sprout from are called nodes and the bases of their leaves are fused around the stem creating a collar.

Horsetail leaves are unusual because they only a single vein. A trait that they share with lycophytes, however the single vein in the *Equisetum* leaves is not believed to be an ancestral trait but rather evolved more recently.

Equisetum species have rhizomes that grow deep below the grounds surface. Some species have cone-like structures called strobili that produce and house the plants spores for reproduction. The tallest species can reach heights of more than 8 m but the majority of species do not grow taller than 2 m.

Distribution of Equisetum

Horsetails enjoy wet environments and some species are even considered semi-aquatic. They are found in all continents except Antarctica and Australia and are also absent from the islands of the South Pacific such as Samoa, Fiji and New Zealand.

Diversity and Taxonomy

There is a total of 15 species in the genus *Equisetum* although many of these species often hybridize with one another. Being ferns, *Equisetum* belong to the division Pteridophyta. They form their own class of ferns known as Equisetopsida and family Equisetaceae.

Reproduction of Equisetum Ferns

Horsetails reproduce using spores rather than seeds. Male and female spores are very similar and are stored in sporangia. Some species store their spores in cone-like structures called strobili.

Equisetum spores have elaters which help to spring spores out of their sporangia once it has split open. The elaters improve the dispersal of spores to new areas by allowing them to get higher and catch more wind. Elaters also enable *Equisetum* spores to travel along the ground.

BRASSICACEAE

Brassicaceae or *Cruciferae* is a medium-sized and economically important family of flowering plants commonly known as the mustards, the crucifers, or the cabbage family. Most are herbaceous plants, some shrubs, with simple, although sometimes deeply incised, alternatingly set leaves without stipules or in leaf rosettes, with terminal inflorescences without bracts, containing flowers with four free sepals, four free alternating petals, two short and four longer free stamens, and a fruit with seeds in rows, divided by a thin wall (or septum).

The family contains 372 genera and 4060 accepted species. The largest genera are *Draba* (440 species), *Erysimum* (261 species), *Lepidium* (234 species), *Cardamine* (233 species), and *Alyssum* (207 species).

The family contains the cruciferous vegetables, including species such as *Brassica oleracea* (e.g., broccoli, cabbage, cauliflower, kale, collards), *Brassica rapa* (turnip, Chinese cabbage, etc.), *Brassica napus* (rapeseed, etc.), *Raphanus sativus* (common radish), *Armoracia rusticana* (horseradish), but also a cut-flower *Matthiola* (stock) and the model organism *Arabidopsis thaliana* (thale cress).

Pieris rapae and other butterflies of the family Pieridae are some of the best-known pests of Brassicaceae species planted as commercial crops. *Trichoplusia ni* (cabbage looper) moth is also becoming increasingly problematic for crucifers due to its resistance to commonly used pest control methods. Some rarer *Pieris* butterflies, such as *Pieris virginiensis*, depend upon native mustards for their survival, in their native habitats. Some non-native mustards, such as garlic mustard, *Alliaria petiolata*, an extremely invasive species in the United States, can be toxic to their larvae.

Species belonging to the Brassicaceae are mostly annual, biennial, or perennial herbaceous plants, some are dwarf shrubs or shrubs, and very few vines. Although generally terrestrial, a few species such as water awlwort live submerged in fresh water. They may have a taproot or a sometimes woody caudex that may have few or many branches, some have thin or tuberous rhizomes, or rarely develop runners. Few species have multi-cellular glands. Hairs consist of one cell and occur in many forms: from simple to forked, star-, tree- or T-shaped, rarely taking the form of a shield or scale. They are never topped by a gland. The stems may be upright, rise up towards the tip, or lie flat, are mostly herbaceous but sometimes woody. Stems carry leaves or the stems may be leafless (in *Caulanthus*), and some species lack stems altogether. The leaves do not have stipules, but there may be a pair of glands at base of leafstalks and flowerstalks. The leaf may be seated or have a leafstalk. The leaf blade is usually simple, entire or dissected, rarely trifoliolate or pinnately compound. A leaf rosette at the base may be present or absent. The leaves along the stem are almost always alternately arranged, rarely apparently opposite. The stomata are of the anisocytic type. The genome size of Brassicaceae compared to that of other Angiosperm families is very small to small (less than 3.425 million base pairs per cell), varying from 150 Mbp in *Arabidopsis thaliana* and *Sphaerocardamum* spp., to 2375 Mbp *Bunias orientalis*. The number of homologous chromosome sets varies from four (n=4) in some *Physaria* and *Stenopetalum* species, five (n=5) in other *Physaria* and *Stenopetalum* species, *Arabidopsis thaliana* and a *Mathiola* species, to seventeen (n=17). About 35% of the species in which chromosomes have been counted have eight sets (n=8). Due to polyploidy, some species may have up to 256 individual chromosomes, with some very high counts in the North American species of *Cardamine*, such as *C. diphylla*. Hybridisation is not unusual in Brassicaceae, especially in *Arabis, Rorippa, Cardamine* and *Boechera*. Hybridisation between species originating in Africa and California, and subsequent polyploidisation is surmised for *Lepidium* species native to Australia and New Zealand.

Inflorescence and Flower

Typical floral diagram of a Brassicaceae (*Erysimum* "Bowles' Mauve").

Flowers may be arranged in racemes, panicles, or corymbs, with pedicels sometimes in the axil of a bract, and few species have flowers that sit individually on flower stems that spring from the axils of rosette leaves. The orientation of the pedicels when fruits are ripe varies dependent on the species. The flowers are bisexual, star symmetrical (zygomorphic in *Iberis* and *Teesdalia*) and the ovary positioned above the other floral parts. Each flower has four free or seldomly merged sepals, the lateral two sometimes with a shallow spur, which are mostly shed after flowering, rarely persistent, may be reflexed, spreading, ascending, or erect, together forming a tube-, bell- or urn-shaped calyx. Each flower has four petals, set alternating with the sepals, although in some species these are rudimentary or absent. They may be differentiated into a blade and a claw or not, and consistently lack basal appendages. The blade is entire or has an indent at the tip, and may sometimes be much smaller than the claws. The mostly six stamens are set in two whorls: usually the two lateral, outer ones are shorter than the four inner stamens, but very rarely the stamens can all have the same length, and very rarely species have different numbers of stamens such as sixteen to twenty four in *Megacarpaea*, four in *Cardamine hirsuta*, and two in *Coronopus*. The filaments are slender and not fused, while the anthers consist of two pollen producing cavities, and open with longitudinal slits. The pollen grains are tricolpate. The receptacle carries a variable number of nectaries, but these are always present opposite the base of the lateral stamens.

Ovary, Fruit and Seed

There is one superior pistil that consists of two carpels that may either sit directly above the base of the stamens or on a stalk. It initially consists of only one cavity but during its further development a thin wall grows that divides the cavity, both placentas and separates the two valves (a so-called false septum). Rarely, there is only one cavity without a septum. The 2–600 ovules are usually along the side margin of the carpels, or rarely at the top. Fruits are capsules that open with two valves, usually towards the top. These are called silique if at least three times longer than wide, or silicle if the length is less than three times the width. The fruit is very variable in its other traits. There may be one persistent style that connects the ovary to the globular or conical stigma, which is undivided or has two spreading or connivent lobes. The variously shaped seeds are usually yellow or brown in color, and arranged in one or two rows in each cavity. The seed leaves are entire or have a notch at the tip. The seed does not contain endosperm.

Differences with Similar Families

Brassicaceae have a bisymmetical corolla (left is mirrored by right, stem-side by out-side, but each quarter is not symmetrical), a septum dividing the fruit, lack stipules and have simple (although sometimes deeply incised) leaves. The sister family Cleomaceae has bilateral symmetrical corollas (left is mirrored by right, but stem-side is different from out-side), stipules and mostly palmately divided leaves, and mostly no septum. Capparaceae generally have a gynophore, sometimes an androgynophore, and a variable number of stamens.

Phytochemistry

Almost all Brassicaceae have C3 carbon fixation. The only exceptions are a few *Moricandia* species, which have a hybrid system between C3 and C4 carbon fixation, C4 fixation being more efficient in drought, high temperature and low nitrate availability. Brassicaceae contain different cocktails of dozens of glucosinolates. They also contain enzymes called myrosinases, that convert the glucosinolates into isothiocyanates, thiocyanates and nitriles, which are toxic to many organisms, and so help guard against herbivory.

Enemies

In New Zealand and Europe the plant is commonly preyed upon by the parasite *Scaptomyza flava*.

Distribution

Brassicaceae can be found almost on the entire land surface of the planet, but the family is absent from Antarctica, and also absent from some areas in the tropics i.e. northeastern Brazil, the Congo basin, Maritime Southeast Asia and tropical Australasia. The area of origin of the family is possibly the Irano-Turanian Region, where approximately 900 species occur in 150 different genera. About 530 of those 900 species are endemics. Next in abundance comes the Mediterranean Region, with around 630 species (290 of which are endemic) in 113 genera. The family is less prominent in the

Saharo-Arabian Region—65 genera, 180 species of which 62 are endemic—and North America (comprising the North American Atlantic Region and the Rocky Mountain Floristic Region)—99 genera, 780 species of which 600 are endemic. South-America has 40 genera containing 340 native species, Southern Africa 15 genera with over 100 species, and Australia and New-Zealand have 19 genera with 114 species between them.

Ecology

Brassicaceae are almost exclusively pollinated by insects. A chemical mechanism in the pollen is active in many species to avoid selfing. Two notable exceptions are exclusive self pollination in closed flowers in *Cardamine chenopodifolia*, and wind pollination in *Pringlea antiscorbutica*. Although it can be pollinated, *Alliaria petiolata* is self-fertile. Most species reproduce sexually through seed, but *Cardamine bulbifera* produces gemmae and in others, such as *Cardamine pentaphyllos*, the coral-like roots easily break into segments, that will grow into separate plants. In some species, such as in the genus *Cardamine*, seed pods open with force and so catapult the seeds quite far. Many of these have sticky seed coats, assisting long distance dispersal by animals, and this may also explain several intercontinental dispersal events in the genus, and its near global distribution. Brassicaceae are common on serpentine and dolomite rich in magnesium. Over a hundred species in the family accumulate heavy metals, particularly zinc and nickel, which is a record percentage. Several *Alyssum* species can accumulate nickel up to 0.3% of their dry weight, and may be useful in soil remediation or even bio-mining.

Brassicaceae contain glucosinolates as well as myrosinases inside their cells. When the cell is damaged, the myrosinases hydrolise the glucosinolates, leading to the synthesis of isothiocyanates, which are compounds toxic to most animals, fungi and bacteria. Some insect herbivores have developed counter adaptations such as rapid absorption of the glucosinates, quick alternative breakdown into non-toxic compounds and avoiding cell damage. In the whites family (Pieridae), one counter mechanisme involves glucosinolate sulphatase, which changes the glucosinolate, so that it cannot be converted to isothiocyanate. A second is that the glucosinates are quickly broken down, forming nitriles. Differences between the mixtures of glucosinolates between species and even within species is large, and individual plants may produce in excess of fifty individual substances. The energy penalty for synthesising all these glucosinolates may be as high as 15% of the total needed to produce a leaf. *Bittercress (Barbarea vulgaris)* also produces triterpenoid saponins. These adaptations and counter adaptations probably have led to extensive diversification in both the Brassicaceae and one of its major pests, the butterfly family Pieridae. A particular cocktail of volatile glucosinates triggers egg-laying in many species. Thus a particular crop can sometimes be protected by planting bittercress as a deadly bait, for the saponins kill the caterpillars, but the butterfly is still lured by the bittercress to lay its egg on the leaves. A moth that feeds on a range of Brassicaceae is the diamondback moth (*Plutella xylostella*). Like the Pieridae, it is capable of converting isothiocyanates into less problematic nitriles. Managing this pest in crops became more complicated after resistance developed against a toxin produced by *Bacillus thuringiensis*, which is used as a wide spectrum biological plant protection against caterpillars. Parasitoid wasps that feed on such insect herbivores are attracted to the chemical compounds released by the plants, and thus are able to locate their prey. The cabbage aphid (*Brevicoryne brassicae*) stores glucosinolates and synthesises its own myrosinases, which may deter its potential predators.

Since its introduction in the 19th century, *Alliaria petiolata* has been shown to be extremely successful as an invasive species in temperate North America due, in part, to its secretion of allelopathic chemicals. These inhibit the germination of most competing plants and kill beneficial soil fungi needed by many plants, such as many tree species, to successfully see their seedlings grow to maturity. The monoculture formation of an herb layer carpet by this plant has been shown to dramatically alter forests, making them wetter, having fewer and fewer trees, and having more vines such as poison ivy, *Toxicodendron radicans*. The overall herb layer biodiversity is also drastically reduced, particularly in terms of sedges and forbs. Research has found that removing 80 percent of the garlic mustard infestation plants did not lead to a particularly significant recovery of that diversity. Instead, it required around 100 percent removal. Given that not one of an estimated 76 species that prey on the plant has been approved for biological control in North America and the variety of mechanisms the plant has to ensure its dominance without them (e.g. high seed production, self-fertility, allelopathy, spring growth that occurs before nearly all native plants, roots that break easily when pulling attempts are made, a complete lack of palatability for herbivores at all life stages, etc.) it is unlikely that such a high level of control can be established and maintained on the whole. It is estimated that adequate control can be achieved with the introduction of two European weevils, including one that is monophagous. However, the USDA's TAG group has blocked these introductions. In addition to being invasive, garlic mustard also is a threat to native North American *Pieris* butterflies such as *Pieris oleracea*, as they preferentially oviposit on it, although it is toxic to their larvae.

Uses

Lunaria annua with dry walls of the fruit.

Smelowskia americana is endemic to the midlatitude mountains of western North America.

This family includes important agricultural crops, among which many vegetables such as cabbage, broccoli, cauliflower, kale, Brussels sprouts, collard greens, Savoy, kohlrabi, and gai lan (*Brassica oleracea*), turnip, napa cabbage, bomdong, bok choy and rapini (*Brassica rapa*), rocket salad/arugula (*Eruca sativa*), garden cress (*Lepidium sativum*), watercress (*Nasturtium officinale*) and radish (*Raphanus*) and a few spices like horseradish (*Armoracia rusticana*), *Brassica*, wasabi (*Eutrema japonicum*), white, Indian and black mustard (*Sinapis alba, Brassica juncea* and *B. nigra* respectively). Vegetable oil is produced from the seeds of several species such as *Brassica napus* (rapeseed oil), perhaps providing the largest volume of vegetable oils of any species. Woad (*Isatis tinctoria*) was used in the past to produce a blue textile dye (indigo), but has largely been replaced by the same substance from unrelated tropical species like *Indigofera tinctoria*.

Brassinosteroids are growing in agricultural and gardening importance.

The Brassicaceae also includes ornamentals, such as species of *Aethionema, Alyssum, Arabis, Aubrieta, Aurinia, Cheiranthus, Erysimum, Hesperis, Iberis, Lobularia, Lunaria, Malcolmia,* and

Matthiola. Honesty (*Lunaria annua*) is cultivated for the decorative value of the translucent remains of the fruits after drying. However, it can be a pest species in areas where it is not native.

The small Eurasian weed *Arabidopsis thaliana* is widely used as model organism in the study of the molecular biology of flowering plants (Angiospermae).

Some species are useful as food plants for Lepidoptera, such as certain wild mustard and cress species, such as *Turritis glabra* and *Boechera laevigata* that are utilized by several North American butterflies. Garlic mustard, *Alliaria petiolata*, is one of the most aggressive and damaging invasive species in North America. Invasive aggressive mustard species are known for being self-fertile, seeding very heavily with small seeds that have a lengthy lifespan coupled with a very high rate of viability and germination, and for being completely unpalatable to both herbivores and insects in areas they are not native to. Garlic mustard is toxic to several rarer North American *Pieris* species.

AMARANTH

Amaranthus, collectively known as amaranth, is a cosmopolitan genus of annual or short-lived perennial plants. Some amaranth species are cultivated as leaf vegetables, pseudocereals, and ornamental plants. Most of the *Amaranthus* species are summer annual weeds and are commonly referred to as pigweed. Catkin-like cymes of densely packed flowers grow in summer or autumn. Approximately 60 species are recognized, with inflorescences and foliage ranging from purple, through red and green to gold. Members of this genus share many characteristics and uses with members of the closely related genus *Celosia*.

Taxonomy

Skull shapes made of amaranth and honey.

Traditional Mexican candy made with amaranth.

Amaranthus shows a wide variety of morphological diversity among and even within certain species. Although the family (Amaranthaceae) is distinctive, the genus has few distinguishing characters among the 70 species included. This complicates taxonomy and *Amaranthus* has generally been considered among systematists as a "difficult" genus.

Formerly, Sauer classified the genus into two subgenera, differentiating only between monoecious and dioecious species: *Acnida* (L.) Aellen ex K.R. Robertson and *Amaranthus*. Although this classification was widely accepted, further infrageneric classification was (and still is) needed to differentiate this widely diverse group.

Currently, *Amaranthus* includes three recognised subgenera and 70 species, although species numbers are questionable due to hybridisation and species concepts. Infrageneric classification focuses on inflorescence, flower characters and whether a species is monoecious/dioecious, as in the Sauer suggested classification. A modified infrageneric classification of *Amaranthus* was published by Mosyakin and Robertson includes three subgenera: *Acnida*, *Amaranthus*, and *Albersia*. The taxonomy is further differentiated by sections within each of the subgenera.

Species

Species include:

- *Amaranthus acanthochiton* – greenstripe
- *Amaranthus acutilobus* – a synonym of *Amaranthus viridis*
- *Amaranthus albus* – white pigweed, tumble pigweed
- *Amaranthus anderssonii*
- *Amaranthus arenicola* – sandhill amaranth
- *Amaranthus australis* – southern amaranth
- *Amaranthus bigelovii* – Bigelow's amaranth
- *Amaranthus blitoides* – mat amaranth, prostrate amaranth, prostrate pigweed
- *Amaranthus blitum* – purple amaranth
- *Amaranthus brownii* – Brown's amaranth
- *Amaranthus californicus* – California amaranth, California pigweed
- *Amaranthus cannabinus* – tidal-marsh amaranth
- *Amaranthus caudatus* – love-lies-bleeding, pendant amaranth, tassel flower, *quilete*
- *Amaranthus chihuahuensis* – Chihuahuan amaranth
- *Amaranthus crassipes* – spreading amaranth
- *Amaranthus crispus* – crispleaf amaranth
- *Amaranthus cruentus* – purple amaranth, red amaranth, Mexican grain amaranth
- *Amaranthus hybridus* – smooth amaranth, smooth pigweed, red amaranth
- *Amaranthus hypochondriacus* – Prince-of-Wales feather, prince's feather
- *Amaranthus interruptus* – Australian amaranth
- *Amaranthus minimus*
- *Amaranthus mitchellii*
- *Amaranthus muricatus* – African amaranth
- *Amaranthus obcordatus* – Trans-Pecos amaranth
- *Amaranthus palmeri* – Palmer's amaranth, Palmer pigweed, careless weed
- *Amaranthus polygonoides* – tropical amaranth
- *Amaranthus powellii* – green amaranth, Powell amaranth, Powell pigweed
- *Amaranthus pringlei* – Pringle's amaranth
- *Amaranthus pumilus* – seaside amaranth
- *Amaranthus retroflexus* – red-root amaranth, redroot pigweed, common amaranth
- Amaranthus saradhiana
- *Amaranthus scleranthoides* – variously *Amaranthus sclerantoides*
- *Amaranthus scleropoides* – bone-bract amaranth
- *Amaranthus spinosus* – spiny amaranth, prickly amaranth, thorny amaranth

- *Amaranthus deflexus* – large-fruit amaranth
- *Amaranthus dubius* – spleen amaranth, *khada sag*
- *Amaranthus fimbriatus* – fringed amaranth, fringed pigweed
- *Amaranthus floridanus* – Florida amaranth
- *Amaranthus furcatus*
- *Amaranthus graecizans*
- *Amaranthus grandiflorus*
- *Amaranthus greggii* – Gregg's amaranth
- *Amaranthus standleyanus*
- *Amaranthus thunbergii* – Thunberg's amaranth
- *Amaranthus torreyi* – Torrey's amaranth
- *Amaranthus tricolor* – Joseph's-coat
- *Amaranthus tuberculatus* – rough-fruit amaranth, tall waterhemp
- *Amaranthus viridis* – slender amaranth, green amaranth
- *Amaranthus watsonii* – Watson's amaranth
- *Amaranthus wrightii* – Wright's amaranth

Nutrition

Uncooked amaranth grain is 12% water, 65% carbohydrates (including 7% dietary fiber), 14% protein, and 7% fat. A 100 grams (3.5 oz) reference amount of uncooked amaranth grain provides 371 calories, and is a rich source (20% or more of the Daily Value, DV) of protein, dietary fiber, pantothenic acid, vitamin B6, folate, and several dietary minerals (table). Uncooked amaranth is particularly rich in manganese (159% DV), phosphorus (80% DV), magnesium (70% DV), iron (59% DV), and selenium (34% DV). Cooking decreases its nutritional value substantially across all nutrients, with only dietary minerals remaining at moderate levels.

Cooked amaranth leaves are a rich source of vitamin A, vitamin C, calcium, and manganese, with moderate levels of folate, iron, magnesium, and potassium. Amaranth does not contain gluten.

Phytochemicals

Amaranth grain contains phytochemicals that are not defined as nutrients and may be antinutrient factors, such as polyphenols, saponins, tannins, and oxalates. These compounds are reduced in content and antinutrient effect by cooking.

Human Uses

Amaranth being roasted in a comal.

Known to the Aztecs as *huāuhtli*, amaranth is thought to have represented up to 80% of their energy consumption before the Spanish conquest. Another important use of amaranth throughout Mesoamerica was in ritual drinks and foods. To this day, amaranth grains are toasted much like popcorn and mixed with honey, molasses, or chocolate to make a treat called *alegría*, meaning "joy" in Spanish. Diego Durán described the festivities for the Aztec god Huitzilopochtli (whose name means "left side of the hummingbird"; hummingbirds feed on amaranth flowers). The Aztec month of Panquetzaliztli (7 December to 26 December) was dedicated to Huitzilopochtli. People decorated their homes and trees with paper flags; ritual races, processions, dances, songs, prayers, and finally human sacrifices were held. This was one of the more important Aztec festivals, and the people prepared for the whole month. They fasted or ate very little; a statue of the god was made out of amaranth seeds and honey, and at the end of the month, it was cut into small pieces so everybody could eat a piece of the god. After the Spanish conquest, cultivation of amaranth was outlawed, while some of the festivities were subsumed into the Christmas celebration.

Amaranth is native to the New World and has been first found in the Old World as part of an archaeological excavation in Narhan, India, dated to 1000-800 B.C.E.

Because of its importance as a symbol of indigenous culture, its palatability, ease of cooking, and a protein that is particularly well-suited to human nutritional needs, interest in amaranth seeds (especially *A. cruentus* and *A. hypochondriacus*) revived in the 1970s. It was recovered in Mexico from wild varieties and is now commercially cultivated. It is a popular snack in Mexico, sometimes mixed with chocolate or puffed rice, and its use has spread to Europe and parts of North America. Amaranth and quinoa are pseudocereals because of their similarities to cereals in flavor and cooking.

Seed

Several species are raised for amaranth "grain" in Asia and the Americas.

Ancient amaranth grains still used include the three species, *Amaranthus caudatus*, *Amaranthus cruentus*, and *Amaranthus hypochondriacus*. Although amaranth was cultivated on a large scale in ancient Mexico, Guatemala, and Peru, nowadays it is only cultivated on a small scale there, along with India, China, Nepal, and other tropical countries; thus, the potential exists for further cultivation in those countries, as well as in the U.S. In a 1977 article in *Science*, amaranth was described as "the crop of the future". It has been proposed as an inexpensive native crop that could be cultivated by indigenous people in rural areas for several reasons:

- A small amount of seed plants a large area (seeding rate 1 kg/ha).

- Yields are high compared to the seeding rate: 1000 kg or more per hectare.

- It is easily harvested and easily processed, post harvest, as there are no hulls to remove.

- Its seeds are a source of protein.

- It has rich content of the dietary minerals, calcium, magnesium, phosphorus, and potassium.

- In cooked and edible forms, amaranth retains adequate content of several dietary minerals.

- It is easy to cook. Boil in water with twice the amount of water as grain by volume (or 2.4 times as much water by weight).

- It grows fast and, in three cultivated species, the large seedheads can weigh up to 1 kg and contain a half-million small seeds.

In the United States, amaranth crop is mostly used for seed production. Most amaranth in American food products starts as a ground flour, blended with wheat or other flours to create cereals, crackers, cookies, bread or other baked products. Despite utilization studies showing that amaranth can be blended with other flours at levels above 50% without affecting functional properties or taste, most commercial products use amaranth only as a minor portion of their ingredients despite them being marketed as "amaranth" products.

Leaves, Roots and Stems

Amaranth species are cultivated and consumed as a leaf vegetable in many parts of the world. Four species of *Amaranthus* are documented as cultivated vegetables in eastern Asia: *Amaranthus cruentus, Amaranthus blitum, Amaranthus dubius*, and *Amaranthus tricolor*.

In Indonesia and Malaysia, leaf amaranth is called *bayam*. In the Philippines, the Ilocano word for the plant is *kalunay*; the Tagalog word for the plant is *kilitis* or *kulitis*. In Uttar Pradesh and Bihar in India, it is called *chaulai* and is a popular green leafy vegetable (referred to in the class of vegetable preparations called *saag*). It is called *chua* in Kumaun area of Uttarakhand, where it is a popular red-green vegetable. In Karnataka in India, it is called *harive*. It is used to prepare curries such as *hulee, palya, majjigay-hulee*, and so on. In Kerala, it is called *cheera* and is consumed by stir-frying the leaves with spices and red chili peppers to make a dish called *cheera thoran*. In Tamil Nadu, it is called *mulaikkira* and is regularly consumed as a favourite dish, where the greens are steamed and mashed with light seasoning of salt, red chili pepper, and cumin. It is called *keerai masial*. In *Andhra Pradesh*, this leaf is added in preparation of a popular *dal* called *thotakura pappu* in (Telugu). In Maharashtra, it is called *shravani maath* and is available in both red and white colour. In Orissa, it is called *khada saga*, it is used to prepare *saga bhaja*, in which the leaf is fried with chili and onions.

In China, the leaves and stems are used as a stir-fry vegetable, or in soups. In Vietnam, it is called *rau dền* and is used to make soup. Two species are popular as edible vegetable in Vietnam: *dền đỏ* (*Amaranthus tricolor*) and *dền cơm* or *dền trắng* (*Amaranthus viridis*).

A traditional food plant in Africa, amaranth has the potential to improve nutrition, boost food security, foster rural development and support sustainable land care.

In Bantu regions of Uganda and western Kenya, it is known as *doodo* or *litoto*. It is also known among the Kalenjin as a drought crop (*chepkerta*). In Nigeria, it is a common vegetable and goes with all Nigerian starch dishes. It is known in Yoruba as *shoko* a short form of *shokoyokoto* (meaning "make the husband fat") or *arowo jeja* (meaning "we have money left over for fish"). In the Caribbean, the leaves are called *bhaji* in Trinidad and *callaloo* in Jamaica, and are sautéed with onions, garlic, and tomatoes, or sometimes used in a soup called pepperpot soup. In Botswana, it is referred to as *morug* and cooked as a staple green vegetable.

In Greece, green amaranth (*A. viridis*) is a popular dish called *vlita* or *vleeta*. It is boiled, then served with olive oil and lemon juice like a salad, sometimes alongside fried fish. Greeks stop harvesting the plant (which also grows wild) when it starts to bloom at the end of August.

In Brazil, green amaranth was, and to a degree still is, often considered an invasive species as all other species of amaranth (except the generally imported *A. caudatus* cultivar), though some have traditionally appreciated it as a leaf vegetable, under the names of *caruru* or *bredo*, which is consumed cooked, generally accompanying the staple food, rice and beans.

Oil

Making up about 5% of the total fatty acids of amaranth, squalene is extracted as a vegetable-based alternative to the more expensive shark oil for use in dietary supplements and cosmetics.

Dyes

The flowers of the 'Hopi Red Dye' amaranth were used by the Hopi (a tribe in the western United States) as the source of a deep red dye. Also a synthetic dye was named "amaranth" for its similarity in color to the natural amaranth pigments known as betalains. This synthetic dye is also known as Red No. 2 in North America and E123 in the European Union.

Ornamentals

A. hypochondriacus (prince's feather) flowering.

The genus also contains several well-known ornamental plants, such as *Amaranthus caudatus* (love-lies-bleeding), a vigorous, hardy annual with dark purplish flowers crowded in handsome drooping spikes. Another Indian annual, *A. hypochondriacus* (prince's feather), has deeply veined, lance-shaped leaves, purple on the under face, and deep crimson flowers densely packed on erect spikes.

Amaranths are recorded as food plants for some Lepidoptera (butterfly and moth) species including the nutmeg moth and various case-bearer moths of the genus *Coleophora*: *C. amaranthella*, *C. enchorda* (feeds exclusively on *Amaranthus*), *C. immortalis* (feeds exclusively on *Amaranthus*), *C. lineapulvella*, and *C. versurella* (recorded on *A. spinosus*).

Ecology

Amaranth weed species have an extended period of germination, rapid growth, and high rates of seed production, and have been causing problems for farmers since the mid-1990s. This is partially due to the reduction in tillage, reduction in herbicidal use and the evolution of herbicidal resistance in several species where herbicides have been applied more often. The following 9 species of *Amaranthus* are considered invasive and noxious weeds in the U.S and Canada: *A. albus, A. blitoides, A. hybridus, A. palmeri, A. powellii, A. retroflexus, A. spinosus, A. tuberculatus,* and *A. viridis.*

A new herbicide-resistant strain of *Amaranthus palmeri* has appeared; it is glyphosate-resistant and so cannot be killed by herbicides using the chemical. Also, this plant can survive in tough conditions. The species *Amaranthus palmeri* (Palmer amaranth) causes the greatest reduction in soybean yields and has the potential to reduce yields by 17-68% in field experiments. Palmer amaranth is among the "top five most troublesome weeds" in the southeast of the United States and has already evolved resistances to dinitroaniline herbicides and acetolactate synthase inhibitors. This makes the proper identification of *Amaranthus* species at the seedling stage essential for agriculturalists. Proper weed control needs to be applied before the species successfully colonizes in the crop field and causes significant yield reductions.

An evolutionary lineage of around 90 species within the genus has acquired the C_4 carbon fixation pathway, which increases their photosynthetic efficiency. This probably occurred in the Miocene.

CHENOPODIUM ALBUM

Chenopodium album is a fast-growing weedy annual plant in the genus *Chenopodium*.

Though cultivated in some regions, the plant is elsewhere considered a weed. Common names include lamb's quarters, melde, goosefoot, manure weed, and fat-hen, though the latter two are also applied to other species of the genus Chenopodium, for which reason it is often distinguished as white goosefoot. It is sometimes also called pigweed. However, pigweed is also a name for several other plants in the family Amaranthaceae; it is used, for example, for the redroot pigweed (*Amaranthus retroflexus*).

Chenopodium album is extensively cultivated and consumed in Northern India as a food crop, and in English texts it may be called by its Hindi name *bathua* or *bathuwa*) (Marathi). It is called *pappukura* in Telugu, *paruppukkirai* in Tamil, *kaduoma* in Kannada, *vastuccira* in Malayalam, and *chakvit* in Konkani.

Distribution

Its native range is obscure due to extensive cultivation, but includes most of Europe, from where Linnaeus described the species in 1753. Plants native in eastern Asia are included under *C. album*, but often differ from European specimens. It is widely naturalised elsewhere, e.g. Africa, Australasia, North America, and Oceania, and now occurs almost everywhere (even, apparently in Antarctica) in soils rich in nitrogen, especially on wasteland.

Botany

It tends to grow upright at first, reaching heights of 10–150 cm (rarely to 3 m), but typically becomes recumbent after flowering (due to the weight of the foliage and seeds) unless supported by other plants. The leaves are alternate and varied in appearance. The first leaves, near the base of the plant, are toothed and roughly diamond-shaped, 3–7 cm long and 3–6 cm broad. The leaves on the upper part of the flowering stems are entire and lanceolate-rhomboid, 1–5 cm long and 0.4–2 cm broad; they are waxy-coated, unwettable and mealy in appearance, with a whitish coat on the underside. The small flowers are radially symmetrical and grow in small cymes on a dense branched inflorescence 10–40 cm long. Further, the flowers are bisexual and female, with five tepals which are mealy on outer surface, and shortly united at the base. There are five stamens.

Taxonomy

Chenopodium album has a very complex taxonomy and has been divided in numerous microspecies, subspecies and varieties, but it is difficult to differentiate between them. The following infraspecific taxa are accepted by the *Flora Europaea*:

- *Chenopodium album* subsp. *album.*
- *Chenopodium album* subsp. *striatum* (Krašan) Murr.
- *Chenopodium album* var. *reticulatum* (Aellen) Uotila.

Published names and synonyms include *C. album* var. *microphyllum*, *C. album* var. *stevensii*, *C. acerifolium*, *C. centrorubrum*, *C. giganteum*, *C. jenissejense*, *C. lanceolatum*, *C. pedunculare* and *C. probstii*.

It also hybridises readily with several other *Chenopodium* species, including *C. berlandieri*, *C. ficifolium*, *C. opulifolium*, *C. strictum* and *C. suecicum*.

Cultivation

Regions where Cultivated

The species are cultivated as a grain or vegetable crop (such as in lieu of spinach), as well as animal feed in Asia and Africa, whereas in Europe and North America, it is commonly regarded as a weed in places such as potato fields, while in Australia it is naturalised in all states and regarded as an environmental weed in New South Wales, Victoria, Western Australia and the Northern Territory.

Potential Impact on Conventional Crops

It is one of the more robust and competitive weeds, capable of producing crop losses of up to 13% in corn, 25% in soybeans, and 48% in sugar beets at an average plant distribution. It may be controlled by dark tillage, rotary hoeing, or flaming when the plants are small. Crop rotation of small grains will suppress an infestation. It is easily controlled with a number of pre-emergence herbicides. Its pollen may contribute to hay fever-like allergies.

Beneficial use in Ecological Pest Control

Chenopodium album is vulnerable to leaf miners, making it a useful trap crop as a companion plant. Growing near other plants, it attracts leaf miners which might otherwise have attacked the crop to be protected. It is a host plant for the beet leafhopper, an insect which transmits curly top virus to beet crops.

Uses and Consumption

Food

Rice and *Chenopodium album* leaf curry with onions and potatoes.

The leaves and young shoots may be eaten as a leaf vegetable, either steamed in its entirety, or cooked like spinach, but should be eaten in moderation due to high levels of oxalic acid. Each plant produces tens of thousands of black seeds. These are high in protein, vitamin A, calcium, phosphorus, and potassium. Quinoa, a closely related species, is grown specifically for its seeds. The Zuni people cook the young plants' greens. Bathua seeds also double up for rice and dal. Napoleon Bonaparte is said to have once relied on bathua seeds to feed his troops during lean times.

Archaeologists analysing carbonized plant remains found in storage pits and ovens at Iron Age, Viking Age, and Roman sites in Europe have found its seeds mixed with conventional grains and even inside the stomachs of Danish bog bodies.

In India, the plant is popularly called *bathua* and found abundantly in the winter season. The leaves and young shoots of this plant are used in dishes such as soups, curries, and paratha-stuffed breads, especially popular in Punjab. The seeds or grains are used in *phambra* or *laa-fi*, gruel-type dishes in Himachal Pradesh, and in mildly alcoholic fermented beverages such as *soora* and *ghanti*.

Animal Feed

As some of the common names suggest, it is also used as feed (both the leaves and the seeds) for chickens and other poultry.

Construction

The juice of this plant is a potent ingredient for a mixture of wall plaster, according to the Samarāṅgaṇa Sūtradhāra, which is a Sanskrit treatise dealing with Śilpaśāstra (Hindu science of art and construction).

GLECHOMA HEDERACEA

Glechoma hederacea (syn. *Nepeta glechoma* Benth., *Nepeta hederacea* (L.) Trevir.) is an aromatic, perennial, evergreen creeper of the mint family Lamiaceae. It is commonly known as ground-ivy, gill-over-the-ground, creeping charlie, alehoof, tunhoof, catsfoot, field balm, and run-away-robin. It is also sometimes known as creeping jenny, but that name more commonly refers to *Lysimachia nummularia*. It is used as a salad green in many countries. European settlers carried it around the world, and it has become a well-established introduced and naturalized plant in a wide variety of localities.

It is considered an aggressive invasive weed of woodlands and lawns in some parts of North America. In the absence of any biological control research conducted by the USDA herbicides are relied upon, despite their drawbacks, particularly for woodland ecosystems. The plant's extensive root system makes it difficult to eradicate by hand-pulling.

Glechoma hederacea can be identified by its round to reniform (kidney or fan shaped), crenate (with round toothed edges) opposed leaves 2–3 cm (0.79–1.18 in) diameter, on 3–6 cm (1.2–2.4 in) long petioles attached to square stems which root at the nodes. The plant spreads either by stolon or seed, making it exceptionally difficult to eradicate. It is a variable species, its size being influenced by environmental conditions, from 5–50 cm (2.0–19.7 in) tall.

Glechoma is sometimes confused with common mallow (*Malva neglecta*), which also has round, lobed leaves; but mallow leaves are attached to the stem at the back of a rounded leaf, where ground ivy has square stems and leaves which are attached in the center of the leaf, more prominent rounded lobes on their edges, attach to the stems in an opposite arrangement, and have a hairy upper surface. In addition, mallow and other creeping plants sometimes confused with ground ivy do not spread from nodes on stems. In addition, ground ivy emits a distinctive odor when damaged, being a member of the mint family.

The flowers of *Glechoma* are bilaterally symmetrical, funnel shaped, blue or bluish-violet to lavender, and grow in opposed clusters of two or three flowers in the leaf axils on the upper part of the stem or near the tip. It usually flowers in the spring.

Glechoma thrives in moist shaded areas, but also tolerates sun very well. It is a common plant in grasslands and wooded areas or wasteland. It also thrives in lawns and around buildings since it survives mowing. Part of the reason for its wide spread is its rhizomatous method of reproduction. It will form dense mats which can take over areas of lawn and woodland and thus is considered an invasive or aggressive weed in suitable climates where it is not native.

Ecological Aspects

A number of wild bees fly upon this plant, including *Anthophora furcata*, *Anthidum manicatum*, *Anthophora plumipes*, *Anthophora quadrimaculata*, *Osmia aurulenta*, *Osmia caerulentes*, and *Osmia uncinata*. The plant is also galled by several insects, including *Rondaniola bursaria* (Lighthouse Gall), *Liposthenes glechomae* or *Liposthenes latreillei* (a gall wasp). Despite its name, it is not related to true ivy (*Hedera*).

Reproduction

Glechoma hederecea is gynodiecious, with genets being either female or hermaphrodite. The females depend upon pollen from hermaphrodites for pollination. Female flowers tend to be smaller than hermaphrodite flowers. There is disagreement among biologists as to whether hermaphrodite flowers can pollinate themselves. The plant spends the winter as either a small ramet or a small rosette. It produces flowers between April and July, which are visited by many types of insects, and can be characterized by a generalized pollination syndrome. Each pollinated flower can produce up to four seeds, which are dispersed by the stem bending over and depositing the ripe seeds in the ground adjacent to the parent plant, although ants may carry the seeds further. The seeds germinate a few days after contact with moisture, although they can be stored dry. Dry storage for a period up to a month is thought to improve the germination rate.

The plant can also reproduce clonally, with the stems bending down to the earth and allowing roots to attach themselves. Single clones can grow several metres across, although precise data is not currently available.

Cultivation and Medicinal and Culinary Uses

Some people consider *Glechoma* to be an attractive garden plant, and it is grown in pots and occasionally as a groundcover. Easily cultivated, it grows well in shaded places. A variegated variety is commercially available; in many areas this is the dominant form which has escaped cultivation and become established as an aggressive, adventitious groundcover.

This species is considered a non-native invasive plant in the United States, and has invaded wild areas, sometimes choking out native wildflowers.

Glechoma was also widely used by the Saxons in brewing ale as flavoring, clarification, and preservative, and later by the English, before the introduction of hops into brewing which changed the ale into beer, in the late 15th century. Thus the brewing-related names for the herb of, alehoof, tunhoof, and gill-over-the-ground.

Glechoma has been used in the cheese-making process as a substitute for animal rennet.

Glechoma hederacea seedling: cot = cotyledons; ga = axillary bud.

Traditional Medicine

Glechoma hederacea has been used in the traditional medicine of Europe going back thousands of years: Galen recommends the plant to treat inflammation of the eyes. John Gerard, an English herbalist, recommended it to treat tinnitus, as well as a "diuretic, astringent, tonic and gentle stimulant. Useful in kidney diseases and for indigestion." It has also been used as a "lung herb". Other traditional uses include as an expectorant, astringent, and to treat bronchitis. In the traditional Austrian medicine the herb has been prescribed for internal application as salad or tea for the treatment of a variety of different conditions including disorders associated with the liver and bile, gastrointestinal tract, respiratory tract, kidneys and urinary tract, fever, and flu.

Safety

Although it has been used as a salad green and in herbal medicines for thousands of years, the safety of *Glechoma hederacea* has not been established scientifically, and there is sufficient evidence to warrant caution with its use because it is toxic to cattle and horses. *Glechoma hederacea* is known to contain terpenoids; terpene-rich volatile oils are known to irritate the gastrointestinal tract and kidneys. The volatile oil also contains pulegone, a chemical also occurring in pennyroyal, that is a known irritant, toxic to the liver, and also an abortifacient. The concentration of volatile oil in *Glechoma* is less than 1/30th that in pennyroyal.

Control

A non-native invasive in North America, *Glechoma* is familiar to a large number of people as a weed, a property it shares with many others of the mint family. It can be a problem in heavy, rich soils with good fertility, high moisture, and low boron content. It thrives particularly well in shady areas where grass does not grow well, such as woodlands, although it can also be a problem in full sun.

Because the plant is stoloniferous and will continue to spread from its roots or bits of stem which reroot, even small infestations resist repeated hand weedings. Like crabgrass, *glechoma's* root has a tough-to-remove ball (un-belied by its delicate wide leaves).

There are no biological control agents to help to reduce its spread in North America. Commercial herbicides containing triclopyr are used to control *glechoma*.

Glechoma is also unusually sensitive to boron, and can be killed by applying borax (sodium tetraborate) in solution. However, borax is toxic to ants and to animals at only slightly higher concentrations, and does not break down in the environment. In addition to adverse long-term effects on soil or ground-water, recent research discounts the very efficacy of borax treatment, primarily because finding the correct concentration for a given area is difficult and the potential for damaging desired plants is high.

JACOBAEA VULGARIS

Jacobaea vulgaris, syn. *Senecio jacobaea*, is a very common wild flower in the family Asteraceae that is native to northern Eurasia, usually in dry, open places, and has also been widely distributed as a weed elsewhere.

Common names include ragwort, common ragwort, stinking willie, tansy ragwort, benweed, St. James-wort, stinking nanny/ninny/willy, staggerwort, dog standard, cankerwort, stammer-wort. In the western United States it is generally known as tansy ragwort, or tansy, though its resemblance to the true tansy is superficial.

Although the plant is often unwanted by landowners because of its toxic effect for cattle and horses, and because it is considered a weed by many, it provides a great deal of nectar for pollinators. It was rated in the top 10 for most nectar production (nectar per unit cover per year) in a UK plants survey conducted by the AgriLand project which is supported by the UK Insect Pollinators Initiative. It also was the top producer of nectar sugar in another study in Britain, with a production per floral unit of (2921 ± 448µg).

The plant is generally considered to be biennial but it has the tendency to exhibit perennial properties under certain cultural conditions (such as when subjected to repeated grazing or mowing). The stems are erect, straight, have no or few hairs, and reach a height of 0.3–2.0 metres (1 ft 0 in–6 ft 7 in). The leaves are pinnately lobed and the end lobe is blunt. The many names that include the word "stinking" (and Mare's Fart) arise because of the unpleasant smell of the leaves. The hermaphrodite flower heads are 1.5–2.5 centimetres (0.59–0.98 in) diameter, and are borne in dense, flat-topped clusters; the florets are bright yellow. It has a long flowering period lasting from June to November (in the Northern Hemisphere).

Pollination is by a wide range of bees, flies and moths and butterflies. Over a season, one plant may produce 2,000 to 2,500 yellow flowers in 20- to 60-headed, flat-topped corymbs. The number of seeds produced may be as large as 75,000 to 120,000, although in its native range in Eurasia very few of these would grow into new plants and research has shown that most seeds do not travel a great distance from the parent plant.

Taxonomy

Two subspecies are accepted:

- *Jacobaea vulgaris* ssp. *vulgaris* - the typical plant, with ray florets present.
- *Jacobaea vulgaris* ssp. *dunensis* - the ray florets are missing.

Distribution

Ragwort is abundant in waste land, waysides and grazing pastures. It can be found along road sides, and grows in all cool and high rainfall areas.

Ragwort is native to the Eurasian continent. In Europe it is widely spread, from Scandinavia to the Mediterranean. In Britain and Ireland it is listed as a weed. In the United States it has been introduced, and is present mainly in the northwest and northeast: California, Idaho, Illinois, Maine, Massachusetts, Michigan, Montana, New Jersey, New York, Oregon, Pennsylvania, and Washington.

In South America it grows in Argentina, in Africa in the north, and on the Asian continent in India and Siberia. It is a widespread weed in New Zealand and Australia. In many Australian states ragwort has been declared a noxious weed. This status requires landholders to remove it from their property, by law. The same applies to New Zealand, where farmers sometimes bring in helicopters to spray their farms if the ragwort is too widespread.

Biological Control

Flowering plant with cinnabar moth caterpillars.

Ragwort is a food plant for the larvae of *Cochylis atricapitana*, *Phycitodes maritima*, and *Phycitodes saxicolais*. Ragwort is best known as the food of caterpillars of the cinnabar moth *Tyria jacobaeae*. They absorb alkaloids from the plant and become distasteful to predators, a fact advertised by the black and yellow warning colours. The red and black, day-flying adult moth is also distasteful to many potential predators. The moth is used as a control for ragwort in countries in which it has been introduced and become a problem, like New Zealand and the western United States. As both larvae and adults are distinctly colored and marked, identification of cinnabars is easy outside of their natural range, and grounds and range keepers can quickly recognize them. In both countries, the tansy ragwort flea beetle (*Longitarsus jacobaeae*) has been introduced to combat the plant. Another beetle, *Longitarsus ganglbaueri*, also feeds on ragwort, but will feed on other plants as well, making it an unsuitable biological control. Another biological control agent introduced in the western United States is the ragwort seed fly, although it is not considered very effective at controlling ragwort. The biological control of ragwort was already used in the 1930s.

Ecological Importance

In the United Kingdom, where the plant is native, ragwort provides a home and food source to at least 77 insect species. Thirty of these species of invertebrate use ragwort exclusively as their food source and there are another 22 species where ragwort forms a significant part of their diet.

Drone fly on ragwort.

Furthermore, English Nature identifies a further 117 species that use ragwort as a nectar source whilst travelling between feeding and breeding sites, or between metapopulations. These consist mainly of solitary bees, hoverflies, moths, and butterflies such as the small copper butterfly (*Lycaena phlaeas*). Pollen is collected by solitary bees.

Besides the fact that ragwort is very attractive to such a vast array of insects, some of these are very rare indeed. Of the 30 species that specifically feed on ragwort alone, seven are officially deemed nationally scarce. A further three species are on the IUCN Red List. In short, ragwort is an exclusive food source for ten rare or threatened insect species, including the cinnabar moth (*Tyria jacobaeae*), the picture winged fly (*Campiglossa malaris*), the scarce clouded knot horn moth (*Homoeosoma nimbella*), and the Sussex emerald moth (*Thalera fimbrialis*). The Sussex Emerald has been labelled a Priority Species in the United Kingdom Biodiversity Action Plan. A priority species is one which is "scarce, threatened and declining". The remainder of the ten threatened species include three species of leaf beetle, another picture-winged fly, and three micro moths. All of these species are Nationally Scarce B, with one leaf beetle categorised as Nationally Scarce A.

The most common of those species that are totally reliant on ragwort for their survival is the cinnabar moth. The cinnabar is a United Kingdom Biodiversity Action Plan Species, its status described as "common and widespread, but rapidly declining". This gives yet more evidence of ragwort's important role in maintaining the country's biodiversity and a vitally important component of the native flora.

Poisonous Effects

Structure of senecionine.

Ragwort contains many different alkaloids, making it poisonous to certain animals. Alkaloids which have been found in the plant confirmed by the WHO report EHC 80 are - jacobine, jaconine, jacozine, otosenine, retrorsine, seneciphylline, senecionine, and senkirkine. There is a strong variation between plants from the same location in distribution between the possible alkaloids and even the absolute amount of alkaloids varies drastically.

Ragwort is of concern to people who keep horses and cattle. In areas of the world where ragwort is a native plant, such as Britain and continental Europe, documented cases of proven poisoning are rare. Horses do not normally eat fresh ragwort due to its bitter taste. The result, if sufficient quantity is consumed, can be irreversible cirrhosis of the liver of a form identified as megalocytosis where cells are abnormally enlarged. Signs that a horse has been poisoned include yellow mucous membranes, depression, and lack of coordination.

There is no definitive test for the poisoning however, since megalocytosis is not a change in the liver which is specific to ragwort poisoning. It is also seen in poisoning by other alkylating agents, such as nitrosamines and aflatoxins. Aflatoxins are a common contaminant formed in feedstuffs by moulds. Research in the United Kingdom has produced results showing megalocytosis, which may be due to various causes, to be a relatively uncommon cause of liver disease in horses.

The danger of ragwort is that the toxin can have a cumulative effect. The alkaloid does not actually accumulate in the liver but a breakdown product can damage DNA and progressively kills cells. About 3-7% of the body weight is sometimes claimed as deadly for horses, but an example in the scientific literature exists of a horse surviving being fed over 20% of its body weight. The effect of low doses is lessened by the destruction of the original alkaloids by the action of bacteria in the digestive tract before they reach the bloodstream. There is no known antidote or cure to poisoning, but examples are known from the scientific literature of horses making a full recovery once consumption has been stopped.

The alkaloids can be absorbed in small quantities through the skin but studies have shown that the absorption is very much less than by ingestion. Also some are in the N-oxide form which only becomes toxic after conversion inside the digestive tract and they will be excreted harmlessly.

Some sensitive individuals can suffer from an allergic reaction because ragwort like many members of the compositae family contains sesquiterpene lactones which can cause compositae dermatitis. These are different from the pyrrolizidine alkaloids which are responsible for the toxic effects.

Honey collected from ragwort has been found to contain small quantities of jacoline, jacobine, jacozine, senecionine, and seneciphylline, but the quantities have been judged as too minute to be of concern.

Control Legislation

Republic of Ireland

In the Republic of Ireland, the Noxious Weeds (Thistle, Ragwort, and Dock) Order 1937, issued under the Noxious Weeds Act 1936, declares ragwort as a noxious weed, requiring landowners to control its growth.

United Kingdom

In the United Kingdom, common ragwort (*Senecio jacobaea*) is one of the five plants named as an *injurious weed* under the provisions of the Weeds Act 1959. The word *injurious* in this context indicates that it could be harmful to agriculture, not that it is dangerous to animals, as all the other *injurious weeds* listed are non-toxic. Under the terms of this Act, a land occupier can be required by the Secretary of State for Environment, Food and Rural Affairs to prevent the spread of the plant. However, the growth of the plant is not made illegal by the Act and there is no statutory obligation for control placed upon landowners in general.

The Ragwort Control Act 2003 provides for a code of practice on ragwort but does not place any further legal responsibilities on landowners to control the plant.

BENEFICIAL WEEDS

A beneficial weed is an invasive plant not generally considered domesticated (however, some plants, such as dandelions, in addition to growing wild, are commercially cultivated) that has some companion plant effect, is edible, contributes to soil health, adds ornamental value, or is otherwise beneficial. Beneficial weeds include many wildflowers, as well as other weeds that are commonly removed or poisoned.

White clover is included in some grass seed mixes, because it is a legume that fertilizes the soil.

Soil Health

Dandelions benefit neighboring plant health by bringing up nutrients and moisture with their deep tap roots.

Although erroneously assumed to compete with neighboring plants for food and moisture, some "weeds" provide the soil with nutrients, either directly or indirectly.

- For example, legumes, such as white clover, if they are colonized by the right bacteria (Rhizobium most often) add nitrogen to the soil through the process of nitrogen fixation, where the bacteria has a symbiotic relationship with its hosts roots, "fixing" atmospheric nitrogen (combining it with oxygen or hydrogen) making the nitrogen plant-available (NH_4 or NO_3).

- Others use deep tap roots to bring up nutrients and moisture from beyond the range of normal plants so that the soil improves in quality over generations of that plant's presence.

- Weeds with strong, widespread roots also introduce organic matter to the earth in the form of those roots, turning hard, dense clay dirt into richer, more fertile soil.

- Some plants like tomatoes and corn will "piggyback" on nearby weeds, allowing their relatively weak root systems to go deeper.

Pest Prevention

Crow garlic, like any allium, masks scents from pest insects,
protecting neighboring plants.

Many weeds protect nearby plants from insect pests.

Some beneficial weeds repel insects and other pests through their smell, for example alliums and wormwood. Some weeds mask a companion plant's scent, or the pheromones of pest insects, as with ground ivy, as well as oregano and other mints.

Some also are unpleasant to small animals and ground insects, because of their spines or other features, keeping them away from an area to be protected.

Trap Crops

Some weeds act as trap crops, distracting pests away from valued plants. Insects often search for target plants by smell, and then land at random on anything green in the area of the scent. If they land on an edible "weed", they will stay there instead of going on to the intended victim. Sometimes, they actively prefer the trap crop.

Host-finding Disruption

Recent studies on host-plant finding have shown that flying pests are far less successful if their host-plants are surrounded by any other plant or even "decoy-plants" made of green plastic, cardboard, or any other green material.

- First, they seek plants by scent. Any "weed" that has a scent reduces the odds of them finding crop plants. Examples are Crow Garlic (wild chives) and Ground Ivy (a form of wild mint), both dramatically masking both plant scent and insect pheromones. They cut down Japanese beetle infestation, and caterpillar infestation, for example cabbage worm, tomato hornworm, and even squash bugs.

- Second, once an insect is near its target, it avoids landing on dirt, but lands on the nearest green thing. Bare earth gardening helps them home in perfectly on the victim crop. But if one is using "green mulch", even grass or clover, the odds are that they will make what's called an "inappropriate landing" on some green thing they don't want. They will then fly a short distance at random, and land on any other green thing. If they fail to accidentally hit the right kind of plant after several tries, they give up.

- If they plan to lay eggs on the crop, weeds provide one more line of defense. Even if they find the right plant, in order to ensure that they didn't hit on a dying plant or falling leaf, they then make short leaf-to-leaf flights before laying eggs. They must land on the "right kind of leaf" enough times in sequence, before they will risk laying their eggs. The more other greenery is nearby, the harder it is for them to remain on target and get enough reinforcement. Enough "inappropriate landings", and they give up, heading elsewhere.

One scientific study said that simply having clover growing nearby cut the odds of cabbage root flies hitting the right plant from 36% to 7%.

Companion Plants

Queen Anne's Lace provides shelter to nearby plants, as well as attracting predatory insects that eat pests like caterpillars, and may boost the productivity of tomato plants.

Many plants can grow intercropped in the same space, because they exist on different levels in the same area, providing ground cover or working as a trellis for each other. This healthier style of horticulture is called forest gardening. Larger plants provide a wind break or shelter from noonday sun for more delicate plants.

Green Mulch

Conversely, some intercropped plants provide living mulch effect, used by inhibiting the growth of any weeds that are actually harmful, and creating a humid, cooler microclimate around nearby plants, stabilizing soil moisture more than they consume it for themselves.

Plants such as ryegrass, red clover, and white clover are examples of "weeds" that are living mulches, often welcomed in horticulture.

Herbicide

Repel plants or fungi, through a chemical means known as allelopathy. Specific other plants can be bothered by a chemical emission through their roots or air, slowing their growth, preventing seed germination, or even killing them.

Beneficial Insects

A common companion plant benefit from many weeds is to attract and provide habitat for beneficial insects or other organisms which benefit plants.

For example, wild umbellifers attract predatory wasps and flies. The adults eat nectar, but they feed common garden pests to their offspring.

Some weeds attract lady beetles or the "good" types of nematode, or provide ground cover for predatory beetles.

Uses for Humans

- Some beneficial weeds, such as lamb's quarters and purslane, are edible and highly nutritional. Dandelions, a widespread invasive weed, were introduced to North America originally because they were considered a staple source of food; they were admired for maturing quickly and spreading vastly.

- A number of weeds have been proposed as natural alternate sources for latex (rubber), including goldenrod, from which the tires were made on the car famously given by Henry Ford to Thomas Edison.

- Cocklebur and stinging nettle have been used for natural dyes and medicinal purposes.

- Some plants seem to subtly improve the flavor of other plants around them, for example, stinging nettle, besides being edible if properly cooked, seems to increase essential oil production in nearby herbs.

Examples:

- Clover is a legume. Like other beans, it hosts bacteria that fix nitrogen in the soil. Its vining nature covers the ground, sheltering more moisture than it consumes, providing a humid, cooler microclimate for surrounding plants as a "green mulch". It also is preferred by rodents over many garden crops, reducing the loss of vegetable crops.

- Dandelions possess a deep, strong tap root that breaks up hard soil, benefiting weaker-rooted plants nearby, and draw up nutrients from deeper than shallower-rooted nearby plants can access. They will also excrete minerals and nitrogen through their roots.

- Crow garlic, the wild chives found in sunny parts of a North American yard, has all of the companion plant benefits of other alliums, including repelling japanese beetles, aphids, and rodents, and being believed to benefit the flavor of solanums like tomatoes and peppers. It can be used as a substitute for garlic in cooking, though it may lend a bitter aftertaste.

- Bishop's lace (Queen Anne's lace) works as a nurse plant for nearby crops like lettuce, shading them from overly intense sunlight and keeping more humidity in the air. It attracts predatory wasps and flies that eat vegetable pests. It has a scientifically tested beneficial effect on nearby tomato plants. When it is young it has an edible root, revealing its relationship to the domesticated carrot.

Daucus Carota

Daucus carota, whose common names include wild carrot, bird's nest, bishop's lace, and Queen Anne's lace (North America), is a white, flowering plant in the family Apiaceae, native to temperate regions of Europe and southwest Asia, and naturalized to North America and Australia.

Domesticated carrots are cultivars of a subspecies, *Daucus carota* subsp. *sativus*.

Queen Anne's lace – *Daucus carota*.

Fruit cluster containing oval fruits with hooked spines.

The wild carrot is a herbaceous, somewhat variable biennial plant that grows between 30 and 60 cm (1 and 2 ft) tall, and is roughly hairy, with a stiff, solid stem. The leaves are tripinnate, finely divided and lacy, and overall triangular in shape. The leaves are bristly and alternate in a pinnate pattern that separates into thin segments. The flowers are small and dull white, clustered in flat, dense umbels. The umbels are terminal and approximately 3–4 inches (8–10 cm) wide. They may be pink in bud and may have a reddish or purple flower in the centre of the umbel. The lower bracts are three-forked or pinnate, which distinguishes the plant from other white-flowered umbellifers. As the seeds develop, the umbel curls up at the edges, becomes more congested, and develops a concave surface. The fruits are oval and flattened, with short styles and hooked spines. The fruit is small, dry and bumpy with protective hairs surrounding it. The fruit of *Daucus carota* has two mericarps, or bicarpellate. The endosperm of the fruit grows before the embryo. The dried umbels detach from the plant, becoming tumbleweeds. The function of the tiny red flower, coloured by anthocyanin, is to attract insects. Wild carrot blooms in summer and fall. It thrives best in sun to partial shade. *Daucus carota* is commonly found along roadsides and in unused fields.

Similar in appearance to the deadly poison hemlock, *D. carota* is distinguished by a mix of tripinnate leaves, fine hairs on its solid green stems and on its leaves, a root that smells like carrots, and occasionally a single dark red flower in the center of the umbel.

Uses

Like the cultivated carrot, the *D. carota* root is edible while young, but it quickly becomes too woody to consume. The flowers are sometimes battered and fried. The leaves are also edible except in large quantities.

Extra caution should be used when collecting *D. carota* because it bears a close resemblance to poison hemlock. In addition, the leaves of the wild carrot may cause phytophotodermatitis, so caution should also be used when handling the plant. It has been used as a method of contraception and an abortifacient for centuries.

If used as a dyestuff, the flowers give a creamy, off-white color.

D. carota, when freshly cut, will draw or change color depending on the color of the water in which it is held. This effect is only visible on the "head" or flower of the plant. Carnations also exhibit this effect. This occurrence is a popular science demonstration in primary grade school.

Beneficial Weed

This beneficial weed can be used as a companion plant to crops. Like most members of the umbellifer family, it attracts wasps to its small flowers in its native land; however, where it has been introduced, it attracts very few wasps. In northeast Wisconsin, when introduced with blueberries it did succeed in attracting butterflies and wasps. This species is also documented to boost tomato plant production when kept nearby, and it can provide a microclimate of cooler, moister air for lettuce, when intercropped with it. However, the states of Iowa, Ohio, Michigan and Washington have listed it as a noxious weed, and it is considered a serious pest in pastures. It persists in the soil seed bank for two to five years.

Taste

Several different factors can cause the root of a carrot to have abnormal metabolites (notably 6-methoxymellin) that can cause a bitter taste in the roots. For example, carrots have a bitterer taste when grown in the presence of apples. Also, ethylene can easily produce stress, causing a bitter taste.

Trap Crop

A trap crop is a plant that attracts agricultural pests, usually insects, away from nearby crops. This form of companion planting can save the main crop from decimation by pests without the use of pesticides. While many trap crops have successfully diverted pests off of focal crops in small scale greenhouse, garden and field experiments, only a small portion of these plants have been shown to reduce pest damage at larger commercial scales. A common explanation for reported trap cropping failures, is that attractive trap plants only protect nearby plants if the insects do not move back into the main crop. In a review of 100 trap cropping examples in 2006, only 10 trap crops were classified as successful at a commercial scale, and in all successful cases, trap cropping was supplemented with management practices that specifically limited insect dispersal from the trap crop back into the main crop.

Usage

Trap crops, when used on an industrial scale, are generally planted at a key time in the pest's life-cycle, and then destroyed before that life-cycle finishes and the pest might have transferred from the trap plants to the main crop.

Examples of trap crops include:

- Alfalfa planted in strips among cotton, to draw away lygus bugs, while castor beans surround the field, or tobacco is planted in strips among it, to protect from the budworm *Heliothis*.

- Rose enthusiasts often plant *Pelargonium* geraniums among their rosebushes because Japanese beetles are drawn to the geraniums, which are toxic to them.

- Chervil is used by gardeners to protect vegetable plants from slugs.

- Rye, sesbania, and sicklepod are used to protect soybeans from corn seeding maggots, stink bugs, and velvet green caterpillars, respectively.

- Mustard and Alfalfa planted near strawberries to attract lygus bugs, a method pioneered by Jim Cochran.

Trap crops can be planted around the circumference of the field to be protected, which is assumed to act as a barrier for entry by pests, or they can be interspersed among the focul crop, for example being planted every ninth row. Planting crops in rows helps facilitate supplemental management practices that prevent insect pest dispersal back into the main field, such as driving a vehicle above the trap crop which then removes insect pests by vacuuming them off of the trap crop row or targeted insecticides, which are only deployed on the trap crop. Even if pesticides are used to control insects on the trap crop, total pesticides are greatly reduced in this scenario over conventional agricultural pesticide applications because they are only deployed on a small portion of the farm (the trap crop). Other strategies that prevent dispersal of insect pests back into the main crop include cutting the trap plants, applying predators to the trap plant that eat the pest, and planting a high ratio of trap plants to other plants.

Operation

Recent studies on host-plant finding have shown that flying pests are far less successful if their host-plants are surrounded by any other plant, or even "decoy-plants" made of green plastic, cardboard or any other green material. The host-plant finding process occurs in three phases.

The first phase is stimulation by odours characteristic to the host-plant. This induces the insect to try to land on the plant it seeks. But insects avoid landing on brown (bare) soil. So if only the host-plant is present, the insects will quasi-systematically find it by landing on the only green thing around. This is called an "appropriate landing". When it does an "inappropriate landing", it flies off to any other nearby patch of green. It eventually leaves the area if there are too many "inappropriate" landings.

The second phase of host-plant finding is for the insect to make short flights from leaf to leaf to assess the plant's overall suitability. The number of leaf-to-leaf flights varies according to the insect species and to the host-plant stimulus received from each leaf. But the insect must accumulate sufficient stimuli from the host-plant to lay eggs; so it must make a certain number of consecutive "appropriate" landings. Hence if it makes an "inappropriate landing", the assessment of that plant is negative and the insect must start the process anew.

Thus it was shown that clover used as a ground cover had the same disruptive effect on eight pest species from four insect orders. An experiment showed that 36% of cabbage root flies laid eggs beside cabbages growing in bare soil (which resulted in no crop), compared with only 7% beside cabbages growing in clover (which allowed a good crop). Also that simple decoys made of green card disrupted appropriate landings just as well as did the live ground cover.

Garden Weeds

Broadleaf Plantain (Plantago Major)

Brought into North America by colonists, plantain often pops up where soil is compacted.

- Nutrient Accumulator: Plantain is said to accumulate calcium, sulfur, magnesium, manganese, iron, and silicon.
- Plantain has edible and medicinal properties.

How to use Plantain in the Garden

Plantain benefits the soil if left to grow and die back on its own. For a tidier garden, cut the leaves back monthly and tuck them under the mulch, or lay them on top of the soil to naturally decompose. Leave the roots intact—the plant will either regrow, or the roots will decay, enriching the soil and attracting beneficial soil organisms.

Broadleaf plantain.

Chickweed (Stellaria Media)

Chickweed shows up in disturbed soil such as garden beds and highly tilled areas, indicating low fertility.

- Nutrient Accumulator: Chickweed is said to accumulate potassium and phosphorus.
- Beneficial Insects: Chickweed attracts pollinators searching for nectar in the spring and early summer.
- Chickweed has edible, lettuce-like greens and medicinal properties.

How to use Chickweed in the Garden

Chickweed will benefit the soil if left to grow and die back on its own. For a tidier garden, cut the plants back monthly and tuck them under the mulch, or lay them on top of the soil to naturally decompose. Leave the roots intact—the plant will either regrow, or the roots will decay, enriching the soil and attracting beneficial soil organisms. Cutting it back will reduce its availability to pollinators.

Common chickweed.

Lamb's Quarters (Chenopodium Album)

The presence of lamb's quarters is common in old farm fields, where chemical fertilizers were used in excess. Over time, these "weeds" will improve the soil quality.

- Nutrient Accumulator: Lamb's quarters' deep roots are said to accumulate nitrogen, phosphorus, potassium, calcium, and manganese while loosening the soil.

- Highly nutritious edible properties when found growing in safe environments. The leaves go for a high price to local chefs.

How to use Lamb's Quarters in the Garden

- Lamb's quarters will benefit the soil if left to grow and die back on their own. However, one plant can set over 75,000 seeds.

- For a tidier garden, cut the plants back monthly so they can't flower, and tuck them under the mulch, or lay them on top of the soil to naturally decompose. Leave the roots intact—the plant will either regrow, or the roots will decay, enriching the soil and attracting beneficial soil organisms.

Lamb's quarters.

White Clover (Trifolium Repens)

White clover voluntarily shows up in nitrogen-lacking, dry fields and lawns that cover hardpan clay soil. Lawns where grass clippings are routinely carted away over time become lacking in nitrogen.

- Nitrogen fixer: Nitrogen is necessary for plant growth, and clover can help transfer airborne nitrogen into the soil to be used by neighboring crops.

- Nutrient Accumulator: Clover is said to accumulate phosphorus.

- Beneficial insects: Clover attracts ladybugs, minute pirate bugs, and pollinators looking for nectar. It provides shelter for parasitoid wasps, spiders, and ground beetles. Clover is a preferred egg-laying site for lacewings.

- White clover has edible flowers.

How to use White Clover in the Garden

Permanent Ground Cover: White clover is often used as a permanent ground cover in orchard areas. It covers and protects soil and the shallow fruit tree roots. In the vegetable garden, white clover is often used in pathways, fertilizing nearby garden soil.

White Clover.

Dandelion (Taraxacum Officinale)

Dandelion is one of the most common and arguably the most beneficial of all weeds. It often shows up in hard-pan clay soils, whether in gardens, old fields, or lawns.

- Nutrient Accumulator: Dandelion's deep roots are said to accumulate potassium, phosphorus, calcium, copper, iron, magnesium, and silicon while loosening the soil.

- Beneficial insects: Dandelion attracts ladybugs and pollinators looking for nectar. It also attracts parasitoid wasps and lacewings.

- Dandelion has edible leaves, roots, and flowers with highly medicinal properties.

How to use Dandelion in the Garden

Dandelion will benefit the soil if left to grow and die back on its own. However, one flower seed head can set over 100 seeds.

For a tidier garden, cut the leaves back monthly and tuck them under the mulch, or lay them on top of the soil to naturally decompose. Leave the roots intact—the plant will either regrow, or the roots will decay, enriching the soil and attracting beneficial soil organisms. Cutting them back will reduce their availability to beneficial insects.

Dandelion.

Benefits of Garden Weeds

Here are some of the things beneficial garden weeds do for us:

Weeds Protect Soil

Weeds are fast growing, so they can quickly cover bare ground to protect it. Their roots hold soil together and keep it from eroding away in the wind or rain. Their presence can indicate the need for mulch to protect soil, i.e. more mulch can often mean fewer weeds.

Weeds may Fertilize Soil

Many weeds are said to accumulate vital nutrients from the subsoil and bring the nutrients into their leaves. As the weed leaves die back, they make a healing medicine (fertilizer) for damaged topsoil. Their presence can indicate the need to enrich your soil with amendments such as worm castings or compost. That's because each time you harvest vegetables, you extract nutrients from the soil.

Weeds Condition Soil

Decaying roots—especially deep taproots—add organic matter to the soil. They provide channels for rain and air to penetrate. Decaying roots also create tunnels for worms and other beneficial soil microbes. They help improve the no-till garden.

Weeds Attract Beneficial Insects

Weeds are usually quick to sprout, yet short-lived. For this reason, they flower frequently in order to set seed for the next generation. The flowering and their dense foliage can attract beneficial insects looking for habitat or nectar.

PARASITIC WEEDS

Parasitic weeds belong to a group of plants which have lost their autotrophic way of life during their developmental process (evolution). They are nutrition specialists to the disadvantage of their host plants. With special organs, the haustoria, they penetrate into the vessels of other plants in order to supply themselves with water and nutrients. Different developmental steps in the process of evolution can be understood by the existence of different degrees of parasitism. Facultative parasites like species within the genus Rhinanthus or Melamyprum represent those at the beginning of the process while obligate parasites represent more advanced plants. Obligate parasites are either hemi-parasites with chlorophyll containing species of the genus *Striga* and *Alectra* (Scrophulariaceae), or holoparasites without chlorophyll, e.g., of the genus *Orobanche* (Orobanchaceae). Instead of being root parasites like species of the genus *Orobanche*, *Striga* and *Alectra*, species of the genus Cuscuta (Cuscutaceae) are obligate shoot parasites.

Parasitic weeds of the families Orobanchaceae (*Aeginetia*, *Orobanche*, broomrape) and Scrophulariaceae (*Alectra*, *Striga*, witchweed) are considered to be among the most serious agricultural pests of economic importance in many parts of the world. The genus *Striga* includes about 40 species, of which 11 species are parasites on agricultural crops. The genus *Orobanche* has more than 100 species but only seven are considered as economically significant.

Geographical Distribution and main Host Plants

Parasitic weeds have evolved specificity to crops and plants in the natural vegetation. *Striga hermonthica* (Del.) Benth., *S. asiatica* (L.) Kuntze and *S. gesnerioides* (L.) Vatke, in the given order, are the most economically important species in the semi-arid to sub-humid tropics. The former two species are almost entirely specific to grasses (cereals) such as sorghum (*Sorghum bicolor* (L.) Moench), maize (*Zea mays* L.), pearl millet (*Pennsisetum americanum* L.), rice (*Oryza sativa* L.), sugar cane (*Saccharum officinarum* L.) and others, while the third one is parasitizing dicot hosts, mainly cowpea (*Vigna unguiculata* (L.) Walp.), tobacco (*Nicotiana tabacum* L.) and sweet potato (*Ipomea batatas* (L.) Lam.) Africa was described as the place of origin of the agriculturally important *Striga* species, particularly the Sudano-Ethiopia region, where also sorghum was postulated to be originated. *S. hermonthica* is widespread in the semi-arid zones of northern tropical Africa and it is also found in the south-western part of the Arabian Peninsula. *S. asiatica*, on the other hand, has a wide distribution in the eastern to southern part of Africa, Asia, Australia and the United States. The third species, *S. gesnerioides*, occurs in Africa, the Arabian Peninsula, the Indian subcontinent, and has been introduced to the United States.

The species of the genus *Alectra* are found mainly in tropical Africa and subtropical southern Africa. *A. sessiliflora*, and *A. fluminensis* are also found in subtropical Asia and tropical and subtropical South America, respectively. *A. vogelii* Benth. is the most important species parasitizing mainly grain legumes in sub-Saharan Africa, which include cowpea, bambara groundnut (*Vigna subterranea* (L.) Verdc.), soybean (*Glycine max* (L.) Merr.), mung bean (*Vigna radiata* (L.) Wilczek), groundnut (*Arachis hypogaea* L.) and common bean (*Phaseolus vulgaris* L.).

The Mediterranean region is considered to be one of the centres of origin of *Orobanche* species. The species are distributed worldwide from temperate climates to the semi-arid tropics. The distribution of *Orobanche crenata* Forsk. is restricted to the Mediterranean regions, the Middle East and East Africa (Ethiopia), while other species have a wider spread. Today, the species *O. crenata*, *O. ramosa* L., *O. aegyptiaca* Pers., *O. cernua* Loefl., *O cumana* Wallr., *O. minor* Sm. and *O. foetida* Poir. are one of the major biotic limiting factors to the production of legumes such as faba bean (*Vicia faba* L.), chickpea (*Cicer arietinum* L.), lentil (*Lens culinaris* Medick.), and to crops of the family Solanaceae [tomato (*Lycopersicon esculentum* Mill.), potato (*Solanum tuberosum* L.), and tobacco (*Nicotiana tabacum* L.)] and Asteraceae, mainly sunflower (*Helianthus annuus* L.).

Life Cycle

The seeds of the root-parasitic weeds vary in their ability to germinate immediately after they have reached maturity. Seeds of *Striga* and *Orobanche* are dormant and require a period of after-ripening or so called post-harvest ripening period, whereas seeds of *Alectra vogelli* can germinate immediately after harvest when germination requirements are met. Seed germination occurs when ripened seeds are preconditioned by exposure to warm moist conditions for several days followed by exogenous chemical signals produced by host roots and some non-hosts (germination stimulant). Upon germination, a germ tube, which is in close proximity to the host roots, elongates towards the root of the host, develops an organ of attachment, the haustorium, which serves as a bridge between the parasite and its host, and deprives it of water, mineral nutrients and carbohydrates, causing drought stress and wilting of the host. Stunted shoot growth, leaf chlorosis and reduced photosynthesis are also phenomena that can be observed on susceptible host plants which contribute to reduction of grain yield. Most of the seeds in the soil will not be reached by the stimulant, but will remain viable for up to 15 years, forming a seed reservoir for the next cropping seasons. The penetration of haustorial cells into host tissue (xylem and phloem system) is carried out mechanically by pressure on the host endodermal cells and by hydrolytic enzymes. Conditioning, germination, parasitic contact (attachment) and penetration are mediated by elegant systems of chemical communication between host and parasite. After several weeks of underground development the parasite emerges above the soil surface and starts to flower and produce seeds after another short period of time. Seed production is prodigious, up to 100 000 seeds or even more can be produced by a single plant and lead to a re-infestation of the field. Thus, if host plants are frequently cultivated, the seed population in the soil increases tremendously and cropping of host plants becomes more and more uneconomical.

Agricultural Significance and Yield Losses

A considerable loss in growth and yield of many food and fodder crops is caused by root-parasitic flowering plants. Globally, *Striga* have a greater impact on human welfare than any other parasitic angiosperms because their hosts are subsistence crops in areas marginal for agriculture. In general, low soil fertility, nitrogen deficiency, well-drained soils, and water stress accentuate the severity of *Striga* damage to the hosts. These are typically the environmental conditions for *Striga*-hosts in the semi-arid to subhumid tropics. Nowadays, *Striga* is considered as the greatest single biotic constraint to food production in Africa, where the livelihood of 300 million people is adversely affected. In infested areas, yield losses associated with *Striga* damage are often significant, ranging from 40-100 percent. Moreover, it is predicted that grain production in Africa is potentially at even increasing risk in the future. This is because several factors that influence the occurrence and may accelerate the future spread and the infestation intensity of *Striga* species in agricultural cropping systems. These include the future adaptation of *Striga* to crops and to a wide ecological amplitude, and a drop in soil fertility in tropical soils. The significant yield reductions result in little or no food at all for millions of subsistence farmers and consequently aggravate hunger and poverty.

Alectra vogelii is a serious pest in cowpea production in Africa. The parasite infection did not decrease cowpea dry matter production, but it significantly altered dry matter partitioning by increasing the proportion of root dry matter. Crop yield losses resulting from *A. vogelii* infestation range from 41 percent to total crop loss of highly susceptible cultivars. The yield reduction is mediated through the delayed onset of flowering, reduced number of flowers and pods, and reduced mass of pods and grain.

The damage caused by the parasites *Orobanche* on field and vegetable crops is significant in the Near East, South and East Europe and in various republics of the former Soviet Union. It causes yield losses ranging from 5-100 percent. For example, in Morocco, the infestation of *O. crenata* in food legumes caused yield losses of 32.7 percent on an average in five provinces in the year 1994, which was equal to a production loss of 14 389 tonnes. As a result of the complete devastation caused by *Orobanche* in many areas, production methods had to be modified and cultivation of some susceptible hosts had to be abandoned.

Control Methods: Possibilities and Constraints

Compared with non-parasitic weeds, the control of parasitic weeds has proved to be exceptionally difficult. The ability of the parasite to produce a tremendously high number of seeds, which can remain viable in the soil for more than ten years, and their intimate physiological interaction with their host plants, are the main difficulties that limit the development of successful control measures that can be accepted and used by subsistence farmers. However, several control methods have been tried for the control of parasitic weeds, including cultural and mechanical (crop rotation, trap and catch cropping, fallowing, hand-pulling, nitrogen fertilization, time and method of planting, intercropping and mixed cropping), physical (solarization), chemical (herbicides, artificial seed germination stimulants, e.g. ethylene, ethephon, strigol), use of resistant varieties, and biological. At on-farm level, the management of parasitic weeds is still unsatisfactory since - with the exception of the use of glyphosate in faba bean to control *O. crenata* - present control methods are not efficient enough to control already the underground development stages of the parasites. At present, the restoration of infested fields can only succeed through the improvement of existing farming systems based on a sound analysis of the parasitic weed problem and the development of a sustainable long-term integrated control programme consisting of the more applicable control approaches that are compatible with existing farming systems and with farmer preference and income. The success of cultural measures becomes evident only in the long run and will not improve yields in the present crop, because of the long underground developmental phase as well as the high seed production and longevity. The income of the subsistence farmers is usually too low to justify the use of highly sophisticated technical inputs such as ethylene to trigger ineffective *Striga* seed germination, as used in North Carolina to eradicate *S. asiatica*, or with soil solarization. In addition to the cost, selectivity, low persistence and availability are major constraints that limit the successful usage of herbicides. In addition, the use of synthetic germination stimulants and application of high dosage of nitrogen fertilizer (more than 80 kg N ha^{-1}, mainly as ammonium sulfate or urea), are not readily applicable in African farming systems. Few resistant lines for some host-parasite associations were reported but resistance is often interfered by the large genetic diversity of the parasites. Recent successes have been achieved in biological control, but it has not led to practical field application owing to the difficulties associated with mass rearing, release, formulation and delivery systems.

Striga

Striga, commonly known as Witchweed, is a genus of parasitic plants that occur naturally in parts of Africa, Asia, and Australia. It is in the family Orobanchaceae. Some species are serious pathogens of cereal crops, with the greatest effects being in savanna agriculture in Africa. It also causes considerable crop losses in other regions, including other tropical and subtropical crops in its native range and in the Americas.

Witchweeds are characterized by bright-green stems and leaves and small, brightly colored and attractive flowers. They are obligate hemiparasites of roots and require a living host for germination and initial development, though they can then survive on their own.

The genus is classified in the family Orobanchaceae, although older classifications place it in the Scrophulariaceae.

The number of species is uncertain, but may exceed 40 by some counts.

Hosts and Symptoms

Although most species of *Striga* are not pathogens that affect human agriculture, some species have devastating effects upon crops, particularly those planted by subsistence farmers. Crops most commonly affected are corn, sorghum, rice and sugarcane. Three species cause the most damage: *Striga asiatica, S. gesnerioides*, and *S. hermonthica*.

Witchweed parasitizes maize, millet, sorghum, sugarcane, rice, legumes, and a range of weedy grasses. It is capable of significantly reducing yields, in some cases wiping out the entire crop.

Host plant symptoms, such as stunting, wilting, and chlorosis, are similar to those seen from severe drought damage, nutrient deficiency, and vascular disease.

Lifecycle

Plant roots with connected Striga plant.

Each plant is capable of producing between 90,000 and 500,000 seeds, which may remain viable in the soil for over 10 years. Most seeds produced are not viable. An annual plant, witchweed overwinters in the seed stage. Its seeds germinate in the presence of host root exudate, and develop haustoria which penetrate host root cells. Host root exudate contain strigolactones, signaling molecules that promote striga seed germination. A bell-like swelling forms where the parasitic roots attach to the roots of the host. The pathogen develops underground, where it may spend the next four to seven weeks before emergence, when it rapidly flowers and produces seeds. Witchweed seeds spread readily via wind and water, and in soil via animal vectors. The chief means of dispersal, however, is through human activity, by means of machinery, tools, and clothing.

Haustorium Development

Once germination is stimulated, the *Striga* seed sends out an initial root to probe the soil for the host root. The initial root secretes an oxidizing enzyme that digests the host root surface, releasing quinones. If the quinone product is within the appropriate concentrations, a haustorium will develop from the initial root. The haustorium grows toward the host root until it makes contact with the root surface, establishing parasitic contact in relatively short order. Within 12 hours of initial haustorium growth, the haustorium recognizes the host root and begins rapid cell division and elongation. The haustorium forms a wedge shape and uses mechanical force and chemical digestion to penetrate the host root, pushing the host cells out of the way. Within 48–72 hours, the haustorium has penetrated the host root cortex. Finger-like structures on the haustorium, called oscula penetrate the host xylem through pits in the membrane. The oscula then swell to secure their position within the xylem membrane. *Striga* sieve tubes develop along with the oscula. Shortly after the host xylem is penetrated, *Striga* sieve tubes develop and approach the host phloem within eight cells. This eight cell layer allows for nonspecific nutrient transport from the host to the *Striga* seedling. Within 24 hours after tapping the host xylem and phloem, the *Striga* cotyledons emerge from the seed.

Environment

Temperatures ranging from 30 to 35 °C (86 to 95 °F) in a moist environment are ideal for germination. Witchweed will not develop in temperatures below 20 °C (68 °F). Agricultural soils with a light texture and low nitrogen levels tend to favor Striga's development. Still, witchweed has demonstrated a wide tolerance for soil types if soil temperatures are favorably high. Seeds have been shown to survive in frozen soil of temperatures as low as –15 °C (5 °F), attesting to their aptitude as overwintering structures.

Soil temperature, air temperature, photoperiod, soil type, and soil nutrient and moisture levels do not greatly deter the development of witchweed. These findings,while limited to the Carolinas in the United States, seem to suggest that the pathogen could successfully infect the massive corn crops of the American Midwest.

Management

Management of witchweed is difficult because the majority of its life cycle takes place below ground. If it is not detected before emergence, it is too late to reduce crop losses. To prevent witchweed from spreading it is necessary to plant uncontaminated seeds and to clean soil and plant debris off of machinery, shoes, clothing, and tools before entering fields. If populations are low, hand weeding before seeds are produced is an option.

Striga in the United States has been controlled through the use of several management strategies, including quarantines imposed on affected areas, control of movement of farm equipment between infected and uninfected areas, herbicide application, and imposed "suicidal germination". For the latter, in fields not yet planted in crops, seeds present in the soil are induced to germinate by injecting ethylene gas, which mimics the natural physiological response tied to host recognition. Because no host roots are available, the seedlings die. Unfortunately, each *Striga* plant can produce tens of thousands of tiny seeds, which may remain dormant in the soil for many years. Thus, such treatments do not remove all seeds from the soil. Moreover, this method is expensive and not generally available to farmers in developing nations of Africa and Asia.

Another method called trap cropping involves planting a species in an infested field that will induce the *Striga* seeds to germinate but will not support attachment of the parasite. This method has been used in sorghum plantations by planting *Celosia argentea* between the sorghum rows. Cotton, sunflower and linseed are also effective trap crops. Planting silverleaf (*Desmodium uncinatum*) inhibits striga seed germination and has worked effectively intercropped with maize.

Increasing nitrogen levels in the soil, growing striga-tolerant varieties, trap-cropping, and planting susceptible crops harvested before witchweed seed is produced, are proven anti-striga tactics. Coating maize seeds with fungi or a herbicide also appears to be a promising approach. An example is TAN222, the "striga-resistant" maize variety which is coated with the systemic herbicide imazapyr, to which it is resistant. Any striga seeds sprouting when this maize is in the seedling stage are poisoned when their haustoria embed in the seedling's roots.

Several sorghum varieties have high levels of resistance in local conditions, including 'N-13', 'Framida', and 'Serena'. 'Buruma', 'Shibe', 'Okoa' and 'Serere 17' millet cultivars are considered to be resistant in Tanzania. Some maize varieties show partial resistance to witchweed, including 'Katumani' in Kenya. In a number of rice cultivars, including some cultivars of NERICA (New Rice for Africa), effective pre- and post- attachment resistance mechanisms have been identified. 'StrigAway' herbicide-resistant, herbicide impregnated maize has been shown to reduce the seed bank of striga by 30% in two seasons.

Importance

Maize, sorghum, and sugarcane crops affected by witchweed in the United States have an estimated value well over $20 billion. Furthermore, witchweed is capable of wiping out an entire crop. In fact, it is so prolific that in 1957 Congress allocated money in an attempt to eradicate witchweed. Because of Striga, the Animal and Plant Health Inspection Service (APHIS) of the U.S. Department of Agriculture established a research station and control methods. Through infestation mapping, quarantine, and control activities such as contaminated seed destruction, the acreage parasitized by witchweed has been reduced 99% since its discovery in the United States. APHIS has even offered cash rewards those who identify and report the weed, and encourages landowners to check their own acreage.

Parasitizing important economic plants, witchweed is one of the most destructive pathogens in Africa. In fact, witchweed affects 40% of Africa's arable savanna region, resulting in up to $13 billion lost every year. *Striga* affects 40 million hectares (98,842,153 acres) of crops in sub-Saharan Africa alone. The witchweed infestation is so bad in parts of Africa, some farmers must relocate every few years. The majority of crops in Africa are grown by subsistence farmers who cannot afford expensive witchweed controls, who therefore suffer much as a result of this pathogen.

Species

Common Crop Parasites

Striga asiatica plant.

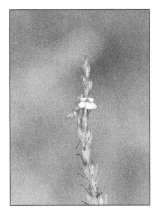

Striga bilabiata.

- *Striga asiatica* has a very wide geographic distribution, from Africa through southern and eastern Asia to Australia. Since the 1950s, it is also known from the United States. This introduction, likely a result of human activity, resulted in an infestation of corn (maize) across many counties in North and South Carolina. The United States Department of Agriculture and state agencies imposed a quarantine on this area to control its spread - a process that was apparently successful.

- *Striga gesnerioides*, cowpea witchweed, as its name implies, is a parasite of cowpea (*Vigna unguiculata*), which is not a grass, but a member of the legume family (Fabaceae or Leguminosae). This species was also accidentally introduced into Florida in the United States, where it was found parasitizing *Indigofera hirsuta* (hairy indigo, another legume).

- *Striga hermonthica* (purple witchweed) is also a parasite that affects grasses, particularly sorghum and pearl millet in sub-Saharan Africa (Senegal to Ethiopia, Democratic Republic of Congo and Tanzania, Angola, Namibia).

Species List

The following are listed as:

- *Striga aequinoctialis* West Africa.

- *Striga angolensis* Angola.

- *Striga angustifolia* East Africa, Asia, Indonesia.

- *Striga asiatica* (Asiatic witchweed) Africa, Arabian peninsula, India, Burma, China, Indonesia, the Philippines, Malaysia, New Guinea, Australia (introduced?), USA (introduced).

- *Striga aspera* Africa.

- *Striga baumannii.*

- *Striga bilabiata* Africa.

- *Striga brachycalyx* Africa.
- *Striga dalzielii* West Africa.
- *Striga densiflora.*
- *Striga elegans* Angola, Malawi, South Africa, Zimbabwe.
- *Striga forbesii* Africa, Madagascar.
- *Striga fulgens.*
- *Striga gesnerioides* (cowpea witchweed) Africa, Arabian peninsula, India, USA (introduced).
- *Striga hermonthica* Senegal to Ethiopia, Democratic Republic of Congo and Tanzania, Angola, Namibia.
- *Striga hirsuta* Madagascar.
- *Striga junodii* South Africa, Mozambique.
- *Striga klingii* West Africa, Nigeria, Ghana, Cameroon, Togo.
- *Striga latericea* East Africa, Ethiopia, Somalia.
- *Striga linearifolia.*
- *Striga macrantha* West Africa, Nigeria, Ivory Coast, Togo.
- *Striga masuria.*
- *Striga passargei* West and Central Africa, Arabian peninsula.
- *Striga primuloides* Ivory Coast, Nigeria.
- *Striga pubiflora* Somalia.
- *Striga strigosa.*

NOXIOUS WEED

A noxious weed, harmful weed or injurious weed is a weed that has been designated by an agricultural or other governing authority as a plant that is injurious to agricultural or horticultural crops, natural habitats or ecosystems, or humans or livestock. Most noxious weeds have been introduced into an ecosystem by ignorance, mismanagement, or accident. Some noxious weeds are native. Typically they are plants that grow aggressively, multiply quickly without natural controls (native herbivores, soil chemistry, etc.), and display adverse effects through contact or ingestion. Noxious weeds are a large problem in many parts of the world, greatly affecting areas of agriculture, forest management, nature reserves, parks and other open space.

Many noxious weeds have come to new regions and countries through contaminated shipments of feed and crop seeds or were intentionally introduced as ornamental plants for horticultural use.

Some "noxious weeds", such as ragwort, produce copious amounts of nectar, valuable for the survival of bees and other pollinators, or other advantages like larval host foods and habitats. Wild parsnip, *Pastinaca sativa*, for instance, provides large tubular stems that some bee species hibernate in, larval food for two different swallowtail butterflies, and other beneficial qualities.

A mature Scotch thistle, an invasive weed.

Types

There are types of noxious weeds that are harmful or poisonous to humans, domesticated grazing animals, and wildlife. Open fields and grazing pastures with disturbed soils and open sunlight are often more susceptible. Protecting grazing animals from toxic weeds in their primary feeding areas is therefore important.

Control

Some guidelines to prevent the spread of noxious weeds are:

1. Avoid driving through noxious weed-infested areas.

2. Avoid transporting or planting seeds and plants that one can't identify.

3. For noxious weeds in flower or with seeds on plants, pulling 'gently' out and placing in a secure closable bag is recommended. Disposal such as hot composting or contained burning is done when safe and practical for the specific plant. Burning poison ivy can be fatal to humans.

4. Using only certified weed-free seeds for crops or gardens.

Maintaining control of noxious weeds is important for the health of habitats, livestock, wildlife and native plants, and of humans of all ages. How to control noxious weeds depends on the surrounding environment and habitats, the weed species, the availability of equipment, labor, supplies, and financial resources. Laws often require that noxious weed control funding from governmental agencies must be used for eradication, invasion prevention, or native habitat and plant community restoration project scopes.

Insects and fungi have long been used as biological controls of different noxious weeds and more recently nematodes have also been used.

AQUATIC WEEDS

Aquatic weeds are the unbated plants which grow and complete their life cycle in water and cause harm to aquatic environment directly and to related eco-environment relatively. Water is one of most important natural resource and in fact basis of all life forms on this planet. Therefore, appropriate management of water from source to its utilization is necessary to sustain the normal function of life. It is one important part of natural resource management. The presence of excessive aquatic vegetation influences the management of water in natural waterways, man made canals and reservoirs which amounts to millions of kilometres/ square kilometers of such water bodies around the world. The area under small tanks and ponds is equally important due to the establishment of many small irrigation schemes and watershed management projects all over the world. For example, India has 1.9 m ha under water in reservoirs and 1.2 m ha under irrigation canals. The area under village ponds and tanks is nearly 2.2 m ha.

Aquatic weeds often reduce the effectiveness of water bodies for fish production. Aquatic weeds can assimilate large quantities of nutrients from the water reducing their availability for planktonic algae. They may also cause reduction in oxygen levels and present gaseous exchange with water resulting in adverse fish production. Although excessive weed growth may provide protective cover in water for small fish growth it may also interfere with fish harvesting.

Dense growth of aquatic weeds may provide ideal habitat for the development of mosquitoes causing malaria, encephality filarasis. These weeds may also serve as vectors for disease causing organisms and can greatly reduce the aesthetic value of water bodies from a recreational point of view.

Aquatic weeds have been found to severely reduce the flow capacity of irrigation canals thereby reducing the availability of water to the farmers field. Aquatic weeds may also damage pumps and turbines in super thermal power stations and hydroelectric power stations influencing electric production and increasing the cost of maintenance of power stations. Many aquatic plants are desirable since they may play temporarily a beneficial role in reducing agricultural, domestic and industrial pollution. Many aquatic weeds may play a useful role of providing continuous supply of phytoplanktons and help fish production.

Aquatic weeds (emergent, floating and submerged) interfere with the static and flow water system. They cause tremendous loss of water from water bodies like lakes and dams through evapo-transpiration. In flowing water system, aquatic weeds impede the flow of water in irrigation canals and drainage channels thereby increasing evaporation damage structures in canals and dams, clog gates, siphons, valves, bridge piers, pump etc. Impediment in flow of water may result in localised floods in neighboring areas. India has the largest canal network in the world where the velocity of flowing water is reduced by about 30 - 40 percent due to the presence of aquatic weeds.

Floating and deep rooted submerged weeds interfere with navigation. Water hyacinth and Alligator weed grow profusely and create dense mats which prevent the movement of boats and at times even large ships.

Village ponds and tanks get infested with floating and submerged weeds which results in reducing the capacity of the water storage and therefore effecting efficient irrigation.

Therefore, considering the losses caused, it is essential to keep aquatic weeds under control in water bodies, flow water systems, ponds and tanks so that these systems can be utilized to best of their efficiency.

Mechanical methods are being practiced at present as use of chemicals is very much restricted due to the difficulty in control on water use for different purposes. Use of bio-agents for weed control is under experimental dissemination and needs further research and refinement in technology for control of aquatic weeds. Within the next two decades bio-agents will be one of the major methods of controlling aquatic weeds, especially the floating ones. Research is also necessary for studying the various factors influencing the aquatic environment and the resultant vegetation. Researchers are envisaging to establish an integrated approach to aquatic weed control using a mix of mechanical and biocidal techniques to control aquatic weeds under specific situations.

Classification of Aquatic Weeds

Aquatic weeds are classified according to various habitats which form their eco-environment and become conducive for their growth, reproduction and dissemination.

Emergent Weeds

These weeds grow in shallow waters and situations existing near the water bodies where water recedes and rises with the seasons or regular releases from a large water body or reservoir. Most of such situations are of permanent in nature where minimum and maximum water levels are consistent. Such situations includes banks of canals, rivers, periphery of water bodies which are mostly in earthen dams, and partly in masonry dams, drainage ditches and water ponds near villages. These weeds may be called semi-aquatic but more appropriately referred to as emergent aquatic weeds. Some examples of the emergent weeds are given below:

Botanical Name	Common Name	Family
Typha angustata	Cattail narrowleaved	Typhaceae
T. latifolia	Cattail common	Typhaceae
T. orientalis	Cattail	Typhaceae
Phragmites communis Trin.	Common reed	Poaceae
P karka	Common reed	Poaceae
P australis	Common reed	Poaceae
Pontederia cordata L	Pickrel weed	Pontedericeae
Commelina benghalensis L	Watergrass	Commelinaceae
Alisma plantago	Water cattail	Alismataceae
Cyperus difformis L	Umbrella plant	Cyperaceae
Ipomea carnea Jacq.	Besharam	Convolvulaceae

There are situations where vast areas of land- remain inundated with water for long periods of time, and may only dry out in severe drought conditions. Such lands are known as marshes or swampy areas. They support a different type of vegetation which may include plants/weeds that are capable of growing under both flooded and saturated conditions. These may include annuals to large trees. Some of these amphibious species are given below:

Temporary Water Situations

Botanical Name	Common Name	Family
Alternanthera philoxeroides(Mart)Griseb	Alligator weed	Amarantheceae
Marsilea minuta L.	Pepper west	Marsileaceae
Meteranthera limosa (SW) Wild	Mud plantain	Pontederaceae
Monochoria vaginalis Presi.	Carpet weed	Pontedereceae
Panicum perpurascens Raddi.	Paragrass	Poaceae
Paspalum fluitans Kunth	Water paspalum	Poaceae

Clay Substratum Situation

Botanical Name	Common Name	Family
Fimbristylis miliacea Vahl	Hoorah grass	Cyperaceae

Floating Mat Situation

Botanical Name	Common Name	Family
Ipomea aquatica fersk	Floating morning glory	Convolvulaceae
Hydrocotyle umbellata L.	Water pennywort	Hydrocolylaceae
Jussiaea repens L	Water primrose	Onagraceae
Ludwigia parviflora	Water purslane	Onagraceae
Trapa bispinosa Roxb.	Water chestnut	Trapaceae

Floating Weeds

These are plants which grow and complete their life cycle in water. They vary in size from single cell (algae) and may grow up to large vascular plants. In case of drying of water bodies most of them give their seeds and other vegetative reproductive organs in base ground lands. These weeds are observed in the surface of the large, deep and shallow depths of water bodies; deep continuous flowing canals; continuously flowing rivers large ponds tanks etc. Some of the weeds in this ecosystem freely float and move long distances, while some of them do float on the water surface but anchor down to soil at the bottom of the water body. These weed species make loss of water through evapotranspiration in addition to impediment caused in flow of water. Therefore, these weeds can be classified in two sub groups. a) Free floating and b) Rooted floating weeds. Examples of common weeds under each sub group are given below:

a) Free Floating Weeds

Botanical Name	Common Name	Family
Eichhornia crassipes (Mart) Solens	Water hyacinth	Pontederiaceae
Salvinia auriculata (Mitch) Syn.	Water fern	Salviniaceae
S molesta	Water fern	Salviniaceae
S natans	Water fern	Salviniaceae
Pistia stratiotes L	Water lettuce	Araceae

Lemna minor	Duck weed	Lemnaceae
Spirodela polyrhiza(L) Schlcid	Giant duck weed	Lemnaceae
Azolla imbricata waxai	Water velvet	Salviniaceae
A pinnata	Water velvet	Salviniaceae
Polygonum amphibium L	Water smart weed	Polygoneaceae

b) Rooted Floating Weeds

Botanical Name	Common Name	Family
Sagittaria guayanensis HBK	Arrowhead	Alismataceae
Ipomea hederacea	Nilkalmi	Convolvulaceae
Nelumbo nucifera G	Lotus	Nymphaceae
Nymphaea alba L	White water lily	Nymphaceae
Nuphar lutea L	Yellow water lily	Nymphaceae
Zannichellia palustris L	Horned pond weed	Zannichelliaceae

Water hyacinth (*Eichhornia crassipes*), a free floating weed.

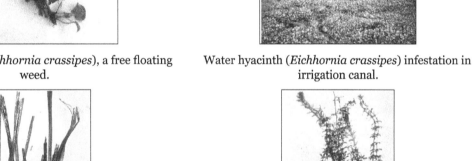

Water hyacinth (*Eichhornia crassipes*) infestation in irrigation canal.

Vallisneria, a submerged aquatic weed.

Hydrilla verticillata, a submerged aquatic weed.

Infestation of *Typha latifolia*, an emersed weed.

Najas, a submerged aquatic weed.

Submerged Weeds

Weed species belonging to this group germinate/ sprout, grow and reproduce beneath the water surface. Their roots and, reproductive organs remain in the soil at the bottom of the water body. These weeds damage the maximum, because they are not visible on the surface and impede the flow of water varying upon the degree of their intensity and growth. Most of these weeds are found in shallow and medium deep water bodies and continuous flowing canals and drainage ditches.

The ecosystem provides situations which allows the growth of algae, filamentous algae, higher algae in shallow water situations and under deep water situations, and thus submerged weeds may be further categorized as a) shallow water submerged weeds, and b) deep water submerged weeds. Commonly occurring weeds in these categories are as follows:

a) Shallow Water Submerged Weeds
Algae

Botanical Name	Common Name	Family
Anabaena spp.	Blue green algae	Nostocaceae
Cladophora spp	Cottonmat type algae	Cladophoraceae
Pithophora spp.	Wet wool type algae	Chloraphyaceae
Spirogyra spp.	Slimy green algae	Chloraphyaceae
Chara zeylanica	Musk grass	Characeae
Nitella hyaline	Stone wort	Characeae

Higher Plants

Botanical Name	Common Name	Family
Najas minor All.	Naiad	Najadaceae
Vallisneria spiralis	Eel weed	Najadaceae
Potamogeton crispus	Curly leaf pond weed	Potamogetonaceae

P. natans L	Broadleaved pond weed	Potamogetonaceae
P. pusillus L.	Small pond weed	Potamogetonaceae
P. nodosus		Potamogetonaceae
P. pectinatus		Potamogetonaceae

b) Deep Water Rooted Submerged Weeds

Botanical Name	Common Name	Family
Myriophyllum spicatum L.	Eurasion water milfoil	Holorhagaceae
Hydrilla verticillata Royle	Hydrilla	Hydrocharitaceae
Elodea Canadensis	Elodea	Hydrocharitaceae
Utricularia flexuosa Vahl.	Bladderwort	Lentibulariaceae

References

- Cirsium-vulgare: pza.sanbi.org, Retrieved 23 January, 2019

- Equisetum, plants- ferns-lycophytes: basicbiology.net, Retrieved 16 May, 2019

- Weeds-you-want-in-garden: tenthacrefarm.com, Retrieved 18 June, 2019

- Brachypodium-sylvaticum: wiki.bugwood.org, Retrieved 02 February, 2019

- Hamdoun, A. M. (1970-09-01). "The Anatomy of Subterranean Structures of Cirsium arvense (L.) Scop". Weed Research. 10 (3): 284–287. doi:10.1111/j.1365-3180.1970.tb00952.x. ISSN 1365-3180

- Rasmussen, Karen; Thyrring, Jakob; Muscarella, Robert; Borchsenius, Finn (16 March 2017). "Climate-change-induced range shifts of three allergenic ragweeds (Ambrosia L.) in Europe and their potential impact on human health". PeerJ. 5: e3104. doi:10.7717/peerj.3104. PMC 5357339. PMID 28321366

Invasive Plants

Invasive plants are the species of plants that are not native to a specific location and can pose serious environmental threats. Tamarix, eichhornia crassipes, reynoutria japonica, cynodon dactylon, ailanthus altissima, cyperus esculentus, purple loosestrife, impatiens glandulifera, etc. are some of the invasive plants. This chapter closely examines these types of invasive plants to provide an extensive understanding of the subject.

INVASIVE PLANT SPECIES

Invasive plants are non-native trees, shrubs, and herbaceous plants that are spread by global trade, human and animal transport and escaping from gardens. They invade forests and block out native plants from growing, which in turn decreases the available habitat for native wildlife. Many invasive plants cannot be used by wildlife for food which puts grazing pressures on the few native plants that remain. They also pose threats to agricultural fields, due to their ability to spread quickly, outcompete crop and forest plants, and deteriorate soil quality. The thick spread of invasive plants makes them costly and time consuming to remove once they have taken hold.

Non-native, invasive plants find their way through a variety of pathways:

- As seeds and weeds in imported nursery plants and soils.

- As misidentified/unknown plants from aquarium keepers, water gardeners, landscapers and friends.

- As whole plants or fragments in ballast water from foreign ships entering our ports.

- As fruits and flowers brought home as souvenirs by travelers.

- On boat trailers, propellers, and dive gear, or in bait wells.

In the past, some species were purposefully introduced to "improve" our natural areas. For example, melaleuca trees (*Melaleuca quinquenervia*) were introduced to Florida from Australia to dry up the Everglades for development. The plan resulted in millions of invasive melaleuca trees covering much of the Everglades. Melaleuca trees have been, and continue to be, removed at a huge expense to taxpayers.

The Chinese tallow tree (*Sapium sebiferum*), was sold widely as an ornamental before it was discovered to be invasive in some of Florida's wetlands. Thousands of Chinese tallow trees have spread on their own in Paynes Prairie Preserve State Park near Gainesville, and much money and labor are now spent controlling them.

Hydrilla (*Hydrilla verticillata*) was introduced as an aquarium plant and sold in pet stores. Hydrilla fluorishes, now infesting tens of thousands of acres in Florida public waters. It has also spread to at least 30 states, as far away as Massachusetts and California. Hydrilla requires continuous management with herbicides and machines. Since 1980, more than $275 million have been spent managing this single species in Florida's public lakes and rivers to conserve their uses and functions.

Flowering plants produce seeds. Some aquatic plants even produce flowers and are pollinated under water. Ferns produce large numbers of tiny spores. Depending on the plant, its location, and other circumstances, plants may spread when:

- Their seeds are dispersed by wind, water, or birds and other animals.

- The plant is fragmented (such as by a boat propeller or mechanical harvester) and the fragments grow into new plants.

- Yard waste or spoil material (from dredging or construction) is dumped else.

- The root system or plant expands and gives rise to new plants.

- Prolific, tiny spores are spread by wind or water, or as a contaminant on clothing, mulch, or landscape material.

Always remove plant matter from boats and trailers after use; some invasive plants can spread
from a small fragment or piece of plant that has broken off and been introduced into a new waterbody.

TAMARIX

The genus *Tamarix* (tamarisk, salt cedar) is composed of about 50–60 species of flowering plants in the family Tamaricaceae, native to drier areas of Eurasia and Africa.

They are evergreen or deciduous shrubs or trees growing to 1–18 m in height and forming dense thickets. The largest, *Tamarix aphylla*, is an evergreen tree that can grow to 18 m tall. They usually grow on saline soils, tolerating up to 15,000 ppm soluble salt and can also tolerate alkaline conditions.

Tamarisks are characterized by slender branches and grey-green foliage. The bark of young branches is smooth and reddish brown. As the plants age, the bark becomes bluish-purple, ridged and furrowed.

The leaves are scale-like, almost like that of junipers, 1–2 mm long, and overlap each other along the stem. They are often encrusted with salt secretions.

The pink to white flowers appear in dense masses on 5–10 cm long spikes at branch tips from March to September, though some species (e.g., *T. aphylla*) tend to flower during the winter.

Reproduction

Tamarix can spread both vegetatively, by adventitious roots or submerged stems, and sexually, by seeds. Each flower can produce thousands of tiny (1 mm diameter) seeds that are contained in a small capsule usually adorned with a tuft of hair that aids in wind dispersal. Seeds can also be dispersed by water. Seedlings require extended periods of soil saturation for establishment. Tamarisk trees are most often propagated by cuttings.

Tamarix species are fire-adapted, and have long tap roots that allow them to intercept deep water tables and exploit natural water resources. They are able to limit competition from other plants by taking up salt from deep ground water, accumulating it in their foliage, and from there depositing it in the surface soil where it builds up concentrations temporarily detrimental to some plants. The salt is washed away during heavy rains.

Tamarix species are used as food plants by the larvae of some Lepidoptera species including *Coleophora asthenella* which feeds exclusively on *T. africana*.

Uses

- Tamarisk species are used as ornamental shrubs, windbreaks, and shade trees: notably *T. ramosissima* and *T. tetrandra*.

- This wood was used by the Saka (combined with wood and ibex horn) to produce tremendously powerful bows hundreds of years before the common era.

- The wood may be used for carpentry or firewood: it is a possible agroforestry species.

- Plans are being made for the tamarisk to play a role in antidesertification programs in China.

- Salt cedars can be planted to mine salts, then be used in the production of fuel and fertilizer (although the latter will be somewhat salty).

Invasive Species

Tamarix ramosissima has naturalized and become a major invasive plant species in parts of the world, such as in the Southwestern United States and Desert Region of California, consuming large amounts of groundwater in riparian and oasis habitats due to the density of its stands. The high salt level in tamarisk infiltrates the soil, preventing other plants from growing, creating a tamarisk-dominant forest with no understory, void of important habitat for pollinators and other native species. Tamarisk forests also tend to burn hotter than most native riparian trees, worsening the fire hazard of acres of uninterrupted tamarisk and their risk to human structures.

Tamarisk eradication projects use a combination of methods, including manual stem cutting followed by application of herbicide to the stump and burning.

KUDZU

Kudzu is a group of plants in the genus *Pueraria*, in the pea family Fabaceae, subfamily Faboideae. They are climbing, coiling, and trailing perennial vines native to much of eastern Asia, Southeast Asia, and some Pacific islands. The name is derived from the Japanese name for the plant East Asian arrowroot. Where these plants are naturalized, they can be invasive and are considered noxious weeds. The plant climbs over trees or shrubs and grows so rapidly that it kills them by heavy shading. The plant is edible, but often sprayed with herbicides.

Flowers of *Pueraria montana*.

Taxonomy and Nomenclature

Kudzu seedpods.

The name kudzu describes one or more species in the genus *Pueraria* that are closely related, and some of them are considered to be varieties rather than full species. The morphological differences between them are subtle; they can breed with each other, and introduced kudzu populations in the United States apparently have ancestry from more than one of the species. They are:

- *P. montana*
- *P. edulis*
- *P. phaseoloides*
- *P. tuberosa*

Propagation

Kudzu spreads by vegetative reproduction via stolons (runners) that root at the nodes to form new plants and by rhizomes. Kudzu also spreads by seeds, which are contained in pods and mature in the autumn, although this is rare. One or two viable seeds are produced per cluster of pods. The hard-coated seeds can remain viable for several years, and can successfully germinate only when soil is persistently soggy for 5-7 days, with temperatures above 20°C (68°F). Once germinated, saplings must be kept in a well-drained medium that retains high moisture. During this stage of growth, kudzu must receive as much sunlight as possible. Kudzu saplings are sensitive to mechanical disturbance, and are damaged by chemical fertilizers. They do not tolerate long periods of shade or high water tables.

Uses

Soil Improvement and Preservation

Kudzu has been used as a form of erosion control and to enhance the soil. As a legume, it increases the nitrogen in the soil by a symbiotic relationship with nitrogen-fixing bacteria. Its deep taproots also transfer valuable minerals from the subsoil to the topsoil, thereby improving the topsoil. In the deforested section of the central Amazon Basin in Brazil, it has been used for improving the soil pore-space in clay latosols, thus freeing even more water for plants than in the soil prior to deforestation.

Animal Feed

Kudzu can be used by grazing animals, as it is high in quality as a forage and palatable to live-stock. It can be grazed until frost and even slightly after. Kudzu had been used in the southern United States specifically to feed goats on land that had limited resources. Kudzu hay typically has a 15–18% crude protein content and over 60% total digestible nutrient value. The quality of the leaves decreases, however, as vine content increases relative to the leaf content. Kudzu also has low forage yields despite its rate of growth, yielding around two to four tons of dry matter per acre annually. It is also difficult to bale due to its vining growth and its slowness in shedding water. This makes it necessary to place kudzu hay under sheltered protection after being baled. Fresh kudzu is readily consumed by all types of grazing animals, but frequent grazing over three to four years can ruin even established stands. Thus, kudzu only serves well as a grazing crop on a temporary basis.

Basketry

Kudzu fiber has long been used for fiber art and basketry. The long runners which propagate the kudzu fields and the larger vines which cover trees make excellent weaving material. Some basketmakers use the material green. Others use it after splitting it in half, allowing it to dry and then rehydrating it using hot water. Both traditional and contemporary basketry artists use kudzu.

Phytochemicals and Uses

Kudzu leaves.

Kudzu contains isoflavones, including puerarin (about 60% of the total isoflavones), daidzein, daidzin (structurally related to genistein), mirificin, and salvianolic acid, among numerous others identified. In traditional Chinese medicine, where it is known as *gé gēn* (gegen), kudzu is considered one of the 50 fundamental herbs thought to have therapeutic effects, although there is no high-quality clinical research to indicate it has any activity or therapeutic use in humans. Adverse effects may occur if kudzu is taken by people with hormone-sensitive cancer or those taking tamoxifen, antidiabetic medications, or methotrexate.

Food

Kuzumochi.

The roots contain starch, which has traditionally been used as a food ingredient in East Asia. In Vietnam, the starch called *bột sắn dây* is flavoured with pomelo oil and then used as a drink in the summer. In Japan, the plant is known as *kuzu* and the starch named *kuzuko*. *Kuzuko* is used in dishes including *kuzumochi, mizu manjū,* and *kuzuyu*. It also serves as a thickener for sauces, and can substitute for cornstarch.

The flowers are used to make a jelly that tastes similar to grape jelly. Roots, flowers, and leaves of kudzu show antioxidant activity that suggests food uses. Nearby bee colonies may forage on kudzu nectar during droughts as a last resort, producing a low-viscosity red or purple honey that tastes of grape jelly or bubblegum.

Herbal Medicine

Kudzu has also been used for centuries in East Asia to make herbal teas and tinctures. Kudzu powder is used in Japan to make an herbal tea called *kuzuyu*. Kakkonto is a herbal drink with its origin in traditional Chinese medicine, intended to be used in situation that the patient has fever, chillness, no perspiration, and the most significant sign, stiff neck and shoulders. It is made from a mixture of sliced fresh ginger, cinnamon twigs, Chinese peony, licorice, jujubes, ephedra, and dried kudzu roots. As the name, which translates literally to "kudzu infusion", implies, kudzu, or *Pueraria lobata*, serves as the main ingredient. Together these plants are used to create a drink containing puerarin, daidzein, paenoflorin, cinnamic acid, glycyrrhizin, ephedrine and gingerol.

Other Uses

Kudzu fiber, known as ko-hemp, is used traditionally to make clothing and paper, and has also been investigated for industrial-scale use.

It may become a valuable asset for the production of cellulosic ethanol. In the Southern United States, kudzu is used to make soaps, lotions, and compost.

Invasive Species

Kudzu growing on trees.

Kudzu plants.

Ecological Damage and Roles

Kudzu's environmental and ecological damage results from its outcompeting other species for a resource. Kudzu competes with native flora for light, and acts to block their access to this vital resource by growing over them and shading them with its leaves. Native plants may then die as a result.

Changes in leaf litter associated with kudzu infestation results in changes to decomposition processes and a 28% reduction in stocks of soil carbon, with potential implications for processes involved in climate change.

United States

Kudzu was introduced from Japan into the United States at the Japanese pavilion in the 1876 Centennial Exposition in Philadelphia. In the 1930s and 1940s, the vine was rebranded as a way for farmers to stop soil erosion. Workers were paid $8 per acre to sow topsoil with the invasive vine. The cultivation covered over one million acres of kudzu. It is now common along roadsides and other disturbed areas throughout most of the southeastern United States as far north as rural areas of Pulaski County, Illinois. Estimates of its rate of spreading differ wildly; it has been described as spreading at the rate of 150,000 acres (610 km²) annually, although in 2015 the United States Forest Service estimated the rate to be only 2,500 acres per year.

Canada

A small patch of Kudzu was discovered in 2009 in Leamington, Ontario, the second warmest growing region of Canada after south coastal British Columbia.

Other Countries

During World War II, kudzu was introduced to Vanuatu and Fiji by United States Armed Forces to serve as camouflage for equipment and has become a major weed.

Kudzu is also becoming a problem in northeastern Australia, and has been seen in Switzerland and in isolated spots in Northern Italy (Lake Maggiore).

In New Zealand, kudzu was declared an "unwanted organism" and was added to the Biosecurity New Zealand register in 2002.

Control

Crown Removal

For successful long-term control of kudzu, destroying the underground system, which can be extremely large and deep, is not necessary. Only killing or removing the kudzu root crown and all rooting runners is needed. The root crown is a fibrous knob of tissue that sits on top of the roots. Crowns form from multiple vine nodes that root to the ground, and range from pea- to basketball-sized. The older the crowns, the deeper they tend to be found in the ground. Nodes and crowns are the source of all kudzu vines, and roots cannot produce vines. If any portion of a root crown remains after attempted removal, the kudzu plant may grow back.

Mechanical methods of control involve cutting off crowns from roots, usually just below ground level. This immediately kills the plant. Cutting off the above-ground vines is not sufficient for an immediate kill. Destroying all removed crown material is necessary. Buried crowns can regenerate into healthy kudzu. Transporting crowns in soil removed from a kudzu infestation is one common way that kudzu unexpectedly spreads and shows up in new locations.

Close mowing every week, regular heavy grazing for many successive years, or repeated cultivation may be effective, as this serves to deplete root reserves. If done in the spring, cutting off vines must be repeated. Regrowth appears to exhaust the plant's stored carbohydrate reserves. Cut kudzu can be fed to livestock, burned, or composted.

The city of Chattanooga, Tennessee, undertook a trial program in 2010 using goats and llamas to graze on the plant. Similar efforts to reduce widespread nuisance kudzu growth have also been undertaken in the cities of Winston-Salem, North Carolina and Tallahassee, Florida.

Prescribed burning is also used on old extensive infestations to remove vegetative cover and promote seed germination for removal or treatment. While fire is not an effective way to kill kudzu, equipment, such as a skid loader, can later remove crowns and thereby kill kudzu with minimal disturbance of soil.

Herbicide

A systemic herbicide, for example, glyphosate, triclopyr, or picloram, can be applied directly on cut stems, which is an effective means of transporting the herbicide into the kudzu's extensive root system. Herbicides can be used after other methods of control, such as mowing, grazing, or burning, which can allow for an easier application of the chemical to the weakened plants. In large-scale forestry infestations, soil-active herbicides have been shown to be highly effective.

After initial herbicidal treatment, follow-up treatments and monitoring are usually necessary, depending on how long the kudzu has been growing in the area. Up to 10 years of supervision may be needed after the initial chemical placement to make sure the plant does not return.

Fungi

Since 1998, the United States Department of Agriculture, Agricultural Research Service (ARS) has experimented with using the fungus *Myrothecium verrucaria* as a biologically based herbicide against kudzu. A diacetylverrucarol spray based on *M. verrucaria* works under a variety of conditions (including the absence of dew), causes minimal injury to many of the other woody plants in kudzu-infested habitats, and takes effect quickly enough that kudzu treated with it in the morning starts showing evidence of damage by midafternoon. Initial formulations of the herbicide produced toxic levels of other trichothecenes as byproducts, though the ARS discovered growing *M. verrucaria* in a fermenter on a liquid diet (instead of a solid) limited or eliminated the problem.

EICHHORNIA CRASSIPES

Eichhornia crassipes, commonly known as common water hyacinth, is an aquatic plant native to the Amazon basin, and is often a highly problematic invasive species outside its native range.

Water hyacinth is a free-floating perennial aquatic plant (or hydrophyte) native to tropical and sub-tropical South America. With broad, thick, glossy, ovate leaves, water hyacinth may rise above the surface of the water as much as 1 meter in height. The leaves are 10–20 cm across on a stem which is floating by means of buoyant bulb like nodules at its base above the water surface. They have long, spongy and bulbous stalks. The feathery, freely hanging roots are purple-black. An erect stalk supports a single spike of 8–15 conspicuously attractive flowers, mostly lavender to pink in colour with six petals. When not in bloom, water hyacinth may be mistaken for frog's-bit (*Limnobium spongia*) or Amazon frogbit (*Limnobium laevigatum*).

One of the fastest growing plants known, water hyacinth reproduces primarily by way of runners or stolons, which eventually form daughter plants. Each plant additionally can produce thousands of seeds each year, and these seeds can remain viable for more than 28 years. Some water hyacinths were found to grow between 2 and 5 metres a day in some sites in Southeast Asia. The common water hyacinth (*Eichhornia crassipes*) are vigorous growers and mats can double in size in two weeks.

In their native range these flowers are pollinated by long tongued bees and they can reproduce both sexually and clonally. The invasiveness of the hyacinth is related to its ability to clone itself and large patches are likely to all be part of the same genetic form.

Water hyacinth have three flower morphs and are termed "tristylous". The flower morphs are named for the length of their pistil: long, medium and short. Tristylous populations are however limited to the native lowland South America range of water hyacinth; in the introduced range, the M-morph prevails, with the L-morph occurring occasionallyand the S-morph is absent altogether. This geographical distribution of the floral morphs indicates that founder events have played a prominent role in the species' worldwide spread.

Habitat and Ecology

Its habitat ranges from tropical desert to subtropical or warm temperate desert to rainforest zones. The temperature tolerance of the water hyacinth is the following; its minimum growth temperature is 12 °C (54 °F); its optimum growth temperature is 25–30 °C (77–86 °F); its maximum growth temperature is 33–35 °C (91–95 °F), and its pH tolerance is estimated at 5.0–7.5. Leaves are killed by frost and plants do not tolerate water temperatures > 34 °C (93 °F). Water hyacinths do not grow where the average salinity is greater than 15% that of sea water (around 5 g salt per kg). In brackish water, its leaves show epinasty and chlorosis, and eventually die. Rafts of harvested water hyacinth have been floated to the sea where it is killed.

Azotobacter chroococcum, a nitrogen-fixing bacteria, is probably concentrated around the bases of the petioles. But the bacteria do not fix nitrogen unless the plant is suffering extreme nitrogen-deficiency.

Fresh plants contain prickly crystals. This plant is reported to contain HCN, alkaloid, and triterpenoid, and may induce itching. Plants sprayed with 2,4-D may accumulate lethal doses of nitrates, and other harmful elements in polluted environments.

Invasive Species

Water hyacinth has been widely introduced in North America, Europe, Asia, Australia, Africa and New Zealand. In many areas it has become an important and pernicious invasive species. In New Zealand it is listed on the National Pest Plant Accord which prevents it from being propagated, distributed or sold. In large water areas such as Louisiana, the Kerala Backwaters in India, Tonlé Sap in Cambodia and Lake Victoria it has become a serious pest. The common water hyacinth has become an invasive plant species on Lake Victoria in Africa after it was introduced into the area in the 1980s.

When not controlled, water hyacinth will cover lakes and ponds entirely; this dramatically affects water flow and blocks sunlight from reaching native aquatic plants which often die. The decay processes depletes dissolved oxygen in the water, often killing fish (or turtles). The plants also create a prime habitat for mosquitos, the classic vectors of disease, and a species of snail known to host a parasitic flatworm which causes schistosomiasis (snail fever). Directly blamed for starving subsistence farmers in Papua New Guinea, water hyacinth remains a major problem where effective control programs are not in place. Water hyacinth is often problematic in man-made ponds if uncontrolled, but can also provide a food source for goldfish, keep water clean and help to provide oxygen.

Water hyacinth often invades bodies of water that have already been affected by human activities. For example, the plants can unbalance natural lifecycles in artificial reservoirs or in eutrophied lakes that receive large amounts of nutrients.

Because of *E. crassipes* invasiveness, several biological control agents have been released to control it, including two weevils (Coleoptera: Curculionidae), *Neochetina bruchi* Hustache and *Neochetina eichhorniae* Warner, and the moth *Niphograpta albiguttalis* (Warren) (Lepidoptera: Pyralidae). *Neochetina eichhorniae* causes "a substantial reduction in water hyacinth production" (in Louisiana); it reduces plant height, weight, root length, and makes the plant produce fewer daughter plants. *N. eichhorniae* was introduced from Argentina to Florida in 1972. A semi-aquatic grasshopper, *Cornops aquaticum*, is being investigated in South Africa as an additional control agent.

United States

The water hyacinth was introduced in 1884 at the World's Fair in New Orleans, also known as the World Cotton Centennial. The plants had been given away as a gift by a group of visiting Japanese people. Soon after, the water hyacinth was choking rivers, killing fish and stopping shipping in

Louisiana, and an estimated 50 kg/m² choked Florida's waterways. There were many attempts to eradicate the species, including one by the U.S. War Department to pour oil over many of the flowers, but none worked. In 1910, a bold solution was put forth by the New Foods Society. Their plan was to import and release hippopotamus from Africa into the rivers and bayous of Louisiana. The hippopotamus would then eat the water hyacinth and also produce meat to solve another serious problem at the time, the American meat crisis.

Known as the American Hippo bill, H.R. 23621 was introduced by Louisiana Congressman Robert Broussard and debated by the Agricultural Committee of the U.S. House of Representatives. The chief collaborators in the New Foods Society and proponents of Broussard's bill were Major Frederick Russell Burnham, the celebrated American Scout, and Captain Fritz Duquesne, a South African Scout who later became a notorious spy for Germany. Presenting before the Agricultural Committee, Burnham made the point that none of the animals that Americans ate, chickens, pigs, cows, sheep, lambs, were native to the U.S.; all had been imported by European settlers centuries before, so why should Americans hesitate to introduce hippopotamus and other large animals into the American diet? Duquesne, who was born and raised in South Africa, further noted that European settlers on that continent commonly included hippopotamus, ostrich, antelope, and other African wildlife in their diets and suffered no ill effects. The American Hippo bill nearly passed, but fell one vote short.

Africa

Water hyacinth at Kisumu Port.

The plant was introduced by Belgian colonists to Rwanda to beautify their holdings. It then advanced by natural means to Lake Victoria where it was first sighted in 1988. There, without any natural enemies, it has become an ecological plague, suffocating the lake, diminishing the fish reservoir, and hurting the local economies. It impedes access to Kisumu and other harbors.

The water hyacinth has also appeared in Ethiopia, where it was first reported in 1965 at the Koka Reservoir and in the Awash River, where the Ethiopian Electric Light and Power Authority has managed to bring it under moderate control at considerable cost of human labor. Other infestations in Ethiopia include many bodies of water in the Gambela Region, the Blue Nile from Lake Tana into Sudan, and Lake Ellen near Alem Tena. By 2018, it has become a serious problem on Lake Tana in Ethiopia.

The water hyacinth is also present on the Shire River in the Liwonde National Park in Malawi.

The water hyacinth invaded Egypt in Muhammad Ali of Egypt's era.

Asia

Water hyacinth has also invaded the Tonlé Sap lake in Cambodia. An Osmose project in Cambodia is trying to fight it by having local people make baskets from it. It was introduced in Bengal in India because of its beautiful flowers and shapes of leaves, but turned out to be an invasive weed draining oxygen from the water bodies and resulted in death of many fish. Fish is a supplement food in Bengal, and because of the fish scarcity in Bengal caused by Eichhornia, the water hyacinth is also called "Terror of Bengal". In Bangladesh, projects have begun to utilize Water hyacinth for the construction of floating vegatable gardens

Europe

In August 2016, the European Union banned any sales of the water hyacinth in the EU.

Control

The three commonly used control methods against water hyacinth infestations are physical, chemical, and biological controls. Each has advantages and drawbacks, although biological control is the best solution in the plant's native environment. The optimum control depends on the specific conditions of each affected location such as the extent of water hyacinth infestation, regional climate, and proximity to human and wildlife.

Chemical Control

Chemical control is the least used out of the three controls of water hyacinth, because of its long-term effects on the environment and human health. The use of herbicides requires strict approval from governmental protection agencies of skilled technician to handle and spray the affected areas. The use of chemical herbicides is only used in case of severe infiltration of water hyacinth. However, the most successful use of herbicides is when it is used for smaller areas of infestation of water hyacinth. This is because in larger areas, more mats of water hyacinths are likely to survive the herbicides and can fragment to further propagate a large area of water hyacinth mats. In addition, it is more cost-effective and less laborious than mechanical control. Yet, it can lead to environmental effects as it can penetrate into the ground water system and can affect not only the hydrological cycle within an ecosystem but also negatively affect the local water system and human health. It is also notable that the use of herbicides is not strictly selective of water hyacinths; keystone species and vital organisms such as microalgae can perish from the toxins and can disrupt fragile food webs.

The chemical regulation of water hyacinths can be done using common herbicides such as 2,4-D, glyphosate, and diquat. The herbicides are sprayed on the water hyacinth leaves and leads to direct changes to the physiology of the plant. The use of the herbicide known as 2,4-D leads to the death of water hyacinth through inhibition of cell growth of new tissue and cellular apoptosis. It can take almost a two-week period before mats of water hyacinth are destroyed with 2, 4-D. Between 75,000 and 150,000 acres (30,000 and 61,000 ha) of water hyacinth and alligator weed are treated annually in Louisiana.

The herbicide known as diquat is a liquid bromide salt that can rapidly penetrate the leaves of the water hyacinth and lead to immediate inactivity of plant cells and cellular processes. For the herbicide glyphosate, it has a lower toxicity than the other herbicides; therefore, it takes longer for the water hyacinth mats to be destroyed (about three weeks). The symptoms include steady wilting of the plants and a yellow discoloration of the plant leaves that eventually leads to plant decay.

Physical Control

Physical control is performed by land-based machines such as bucket cranes, draglines, or boom or by water based machinery such as aquatic weed harvesters, dredges, or vegetation shredder. Mechanical removal is seen as the best short-term solution to the proliferation of the plant. A project on Lake Victoria in Africa used various pieces of equipment to chop, collect, and dispose of 1,500 hectares (3,700 acres) of water hyacinth in a 12-month period. It is, however, costly and requires the use of both land and water vehicles, but it took many years for the lake to become in poor condition and reclamation will be a continual process.

It can have an annual cost from $6 million to $20 million and is only considered a short-term solution to a long-term problem. Another disadvantage with mechanical harvesting is that it can lead to further fragmentation of water hyacinths when the plants are broken up by spinning cutters of the plant-harvesting machinery. The fragments of water hyacinth that are left behind in the water can easily reproduce asexually and cause another infestation.

However, transportation and disposal of the harvested water hyacinth is a challenge because the vegetation is heavy in weight. The harvested water hyacinth can pose a health risk to humans because of the plant's propensity for absorbing contaminants, and it is considered toxic to humans. Furthermore, the practice of mechanical harvesting is not effective in large-scale infestations of the water hyacinth, because this aquatic invasive species grows much more rapidly than it can be eliminated. Only one to two acres ($\frac{1}{2}$ to 1 ha) of water hyacinth can be mechanically harvested daily because of the vast amounts of water hyacinths in the environment. Therefore, the process is very time-intensive.

Biological Control

In 2010 the insect *Megamelus scutellaris* was released by the
Agricultural Research Service as a biological control for water hyacinth.

As chemical and mechanical removal is often too expensive, polluting, and ineffective, researchers have turned to biological control agents to deal with water hyacinth. The effort began in the 1970s when USDA researchers released into the United States three species of weevil known to feed on water hyacinth, *Neochetina bruchi*, *N. eichhorniae*, and the water hyacinth borer *Sameodes albiguttalis*. The weevil species were introduced into the Gulf Coast states, such as Louisiana, Texas, and Florida, where thousands of acres were infested by water hyacinth. It was found that a decade later in the 1980s that there was a decrease in water hyacinth mats by as much as 33%. However, because the life cycle of the weevils is ninety days, it puts a limitation on the use of biological predation to efficiently suppress water hyacinth growth. These organisms regulate water hyacinth by limiting water hyacinth size, its vegetative propagation, and seed production. They also carry microorganisms that can be pathological to the water hyacinth. These weevils eat stem tissue, which results in a loss of buoyancy for the plant, which will eventually sink. Although meeting with limited success, the weevils have since been released in more than 20 other countries. However, the most effective control method remains the control of excessive nutrients and prevention of the spread of this species.

In May 2010, the USDA's Agricultural Research Service released *Megamelus scutellaris* as an additional biological control insect for the invasive water hyacinth species. *Megamelus scutellaris* is a small planthopper insect native to Argentina. Researchers have been studying the effects of the biological control agent in extensive host-range studies since 2006 and concluded that the insect is highly host-specific and will not pose a threat to any other plant population other than the targeted water hyacinth. Researchers also hope that this biological control will be more resilient than existing biological controls and the herbicides that are already in place to combat the invasive water hyacinth.

Another insect being considered as a biological control agent is the semi-aquatic grasshopper *Cornops aquaticum*. This insect is specific to the water hyacinth and its family, and besides feeding on the plant, it introduces a secondary pathogenic infestation. This grasshopper has been introduced into South Africa in controlled trials.

Uses

Bioenergy

Because of its extremely high rate of development, *Eichhornia crassipes* is an excellent source of biomass. One hectare (2.5 acres) of standing crop thus produces more than 70,000 m³/ha (1,000,000 cu ft/acre) of biogas (70% CH_4, 30% CO_2). According to Curtis and Duke, one kg (2.2 lb) of dry matter can yield 370 litres (13 cu ft) of biogas, giving a heating value of 22,000 kJ/m³ (590 Btu/cu ft) compared to pure methane (895 Btu/ft³).

Wolverton and McDonald report approximately 0.2 m³/kg (3 cu ft/lb) methane, indicating biomass requirements of 350 t/ha (160 short ton/acre) to attain the 70,000 m³/ha (1,000,000 cu ft/acre) yield projected by the National Academy of Sciences (Washington). Ueki and Kobayashi mention more than 200 t/ha (90 short ton/acre) per year. Reddy and Tucker found an experimental maximum of more than ½ tonne per hectare (¼ short ton/acre) per day.

Bengali farmers collect and pile up these plants to dry at the onset of the cold season; they then use the dry water hyacinths as fuel. The ashes are used as fertilizer. In India, one tonne (1.1 short tons) of dried water hyacinth yields about 50 liters ethanol and 200 kg residual fiber (7,700 Btu). Bacterial fermentation of one tonne (1.1 short tons) yields 26,500 ft³ gas (600 Btu) with 51.6% methane (CH_4), 25.4% hydrogen (H_2), 22.1% carbon dioxide (CO_2), and 1.2% oxygen (O_2). Gasification of one tonne (1.1 short tons) dry matter by air and steam at high temperatures (800 °C or 1,500 °F) gives about 40,000 ft³ (1,100 m³) natural gas (143 Btu/ft³) containing 16.6% H_2, 4.8% CH_4, 21.7% CO (carbon monoxide), 4.1% CO_2, and 52.8% N_2 (nitrogen). The high moisture content of water hyacinth, adding so much to handling costs, tends to limit commercial ventures. A continuous, hydraulic production system could be designed, which would provide a better utilization of capital investments than in conventional agriculture, which is essentially a batch operation.

The labour involved in harvesting water hyacinth can be greatly reduced by locating collection sites and processors on impoundments that take advantage of prevailing winds. Wastewater treatment systems could also favourably be added to this operation. The harvested biomass would then be converted to ethanol, biogas, hydrogen, gaseous nitrogen, and/or fertilizer. The byproduct water can be used to irrigate nearby cropland.

Phytoremediation and Waste Water Treatment

The roots of *Eichhornia crassipes* naturally absorb pollutants, including lead, mercury, and strontium-90, as well as some organic compounds believed to be carcinogenic, in concentrations 10,000 times that in the surrounding water. Water hyacinths can be cultivated for waste water treatment (especially dairy waste water).

Water hyacinth is reported for its efficiency to remove about 60–80% nitrogen and about 69% of potassium from water. The roots of water hyacinth were found to remove particulate matter and nitrogen in a natural shallow eutrophicated wetland.

Edibility

The plant is used as a carotene-rich table vegetable in Taiwan. Javanese sometimes cook and eat the green parts and inflorescence. Vietnamese also cook the plant and sometimes add its young leaves and flower to their salad.

Medicinal Use

In Kedah (Malaysia), the flowers are used for medicating the skin of horses. The species is a "tonic".

Potential as Bioherbicidal Agent

Water hyacinth leaf extract has been shown to exhibit phytotoxicity against another invasive weed *Mimosa pigra*. The extract inhibited the germination of *Mimosa pigra* seeds in addition to suppressing the root growth of the seedlings. Biochemical data suggested that the inhibitory effects may be mediated by enhanced hydrogen peroxide production, inhibition of soluble peroxidase activity, and stimulation of cell wall-bound peroxidase activity in the root tissues of *Mimosa pigra*.

Other Uses

In East Africa, water hyacinths from Lake Victoria are used to make furniture, handbags and rope. The plant is also used as animal feed and organic fertilizer although there is controversy stemming from the high alkaline pH value of the fertilizer. Though a study found water hyacinths of very limited use for paper production, they are nonetheless being used for paper production on a small scale.

American-Nigerian Achenyo Idachaba has won an award for showing how this plant can be exploited for profit in Nigeria.

In places where water hyacinth is invasive, overabundant, and in need of clearing away, these traits make it free for the harvesting, which makes it very useful as a source of organic matter for composting in organic farming in those locales, provided that the composting method properly handles it. As an aquatic plant, it requires most of the same composting principles as the seaweed that is composted close to sea coasts.

In Bangladesh, farmers in the southwestern region cultivate vegetables on the dried mass of water hyacinth. As a large portion of cultivable land goes under water for months during monsoon in this low-lying region, farmers have grown this method for many decades now. The method of this agriculture is known as 'Dhape chash'.

The plant can also screen heavy metals and various other toxins from contaminated water.

CHRYSANTHEMOIDES MONILIFERA

Chrysanthemoides monilifera is an evergreen flowering shrub or small tree of the Asteraceae (daisy) family that is native to South Africa, such as the Cape Flats Dune Strandveld habitat. Most subspecies have woolly, dull, serrate, oval leaves, but the subspecies *rotundata* has glossy round leaves. Subspecies are known as boneseed and bitou bush in Australasia, or bietou, tick berry, bosluisbessie, or weskusbietou in South Africa. The plant has become a major environmental weed and invasive species in Australia and New Zealand.

Taxonomy

Chrysanthemoides monilifera is one of two members of the genus *Chrysanthemoides*, the other being *Chrysanthemoides incana*.

C. monilifera has six recognized subspecies:

- ssp. *canescens*
- ssp. *monilifera*
- ssp. *pisifera*
- ssp. *rotundata*
- ssp. *septentrionalis*
- ssp. *subcanescens*

In Australia, *C. m.* ssp. *monilifera* is known by the common name 'boneseed', while *C. m.* ssp. *rotundata* is known by the common name 'bitou bush'. In New Zealand subspecies are not distinguished and *C. monilifera* is known simply as 'boneseed'.

C. monilifera was first described by Carl Linnaeus in 1754 under the name *Osteospermum monilifera*, however it was given its current binomial name in 1943 by T. Norlindh.

The species name *monilifera* comes from the *monile*, meaning necklace or collar, referring to the shiny fruit arranged around the flowers like a necklace.

Boneseed is a perennial, woody, upright shrub, growing to 3 m (9.8 ft), although occasionally taller. It is a member of the *Asteraceae* (daisy) family and has showy, bright yellow flowers in swirls of 5-8 'petals' (ray florets) up to 30 mm (1.2 in) in diameter. Fruit are berry-like, spherical at around 8 mm in diameter, and turn dark-brown to black with a bone-coloured seed inside of 6–7 mm diameter. Leaves are 2–6 cm (0.79–2.36 in) long by 1.5–5 cm (0.59–1.97 in) wide, oval tapering to the base with irregularly serrate margins.

Bitou bush can be distinguished from boneseed in part due to its more rounded sprawling habit to 1.5–2 m (4.9–6.6 ft), less noticeably toothy leaf margins and seeds that are egg-like rather than spherical.

Both boneseed and bitou bush hybridise readily, however, so examples of plants demonstrating a fusion of traits is possible.

C. monilifera has been shown to need pollinators in order to reproduce.

Distribution and Habitat

Chrysanthemoides monilifera occurs naturally in coastal areas of South Africa, reaching into southern Namibia and Mozambique. The most widespread subspecies in South Africa is *pisifera*. Subspecies *rotundata* is concentrated along the eastern coast of South Africa from its southern tip through to the Mozambique border. Subspecies *monilifera* is concentrated around Cape Town and the Cape Peninsula on South Africa's south western coast.

Impact

In Australia, *C. monilifera* has been particularly successful in invading natural bushland. In part, this is due to the species' ability to establish on relatively nutrient-poor soils and in areas exposed to salt such as coastlines, as well as the ability of the seeds to germinate readily. Disturbances such as fire can assist *C. monilifera* to spread as the plant produces a large amount of seed that can persist in the soil seed bank for 10 years or more, and this reserve in turn enables the species to quickly recolonize a burnt area.

An individual plant can produce 50,000 seeds a year, about 60% of which are viable. Once germinated, seedlings grow vigorously with dense, bushy growth. This lush growth shades out and displaces slower growing native species that might otherwise occupy the same ecological niche. Rapid, vigorous growth also means that *C. monilifera* is capable of flowering and setting seed within 12–18 months, making it extremely persistent even in situations where disturbance or regular management activity is common.

Once established, the plant's shallow root system enables it to absorb moisture after light rain before the moisture reaches the roots of more deeply rooted species further limiting opportunities for slower growing species to establish and out-compete *C. monilifera* over time. Furthermore, outside of Southern Africa the plant has few local, indigenous pathogens or predators to control its growth also reducing the potential for gaps to emerge that might provide opportunities for other species to reestablish. The net consequence of *C. monilifera*'s growth characteristics is that outside of its natural ecosystem it can ultimately form large, dense, unhealthy stands of a single species with extraordinarily poor biodiversity.

The plant can extend its existing range in a variety of ways. Its fruit is attractive to birds, rabbits, other animals and even some insects such as ants, and because seeds are tough and difficult to digest they will often be dispersed in animal droppings. Seeds can also spread on vehicles and equipment, in contaminated soil, in garden waste, along water drainage lines and deliberately by human intervention.

C. monilifera, unlike many other weed species, is not generally considered to be a problem for agricultural productivity due to its sensitivity to trampling as well as being readily grazed by stock.

Control

C. monilifera is potentially susceptible to a range of control strategies, however Burgman and Lindenmayer recommended that the strategy chosen be responsive to the local situation and available resources. Due to its relatively shallow root system removal by hand is an ideal method of control. Where manual removal is impractical many common herbicides can be used, in which case the herbicide is commonly applied directly to the wood of the plant via a cut notch or at the end of a pruned stump. Mechanical removal of *C. monilifera* by tractor or other machinery can also be effective, however this can be extremely indiscriminate and is only recommended in areas of poor environmental values and minimal erosion risk.

Another method of control available is the use of controlled burns, however there are risks associated with this method. Principally, *C. monilifera* has higher moisture levels than many Australian indigenous species and therefore in Australia a higher than normal intensity fire is required for burns to be effective. This can in turn have detrimental impacts on indigenous vegetation evolved in response to more frequent, lower intensity fire. Furthermore, fire can trigger regermination from the extensive *C. monilifera* seed bank, potentially worsening the situation unless a program is implemented to monitor and control *C. monilifera* seedlings following the burn. If *C. monilifera* seedlings are removed, however, this can be extremely effective at exhausting the seed bank and minimising the chances of reinfestation.

Various methods of biological control have been attempted, particularly the introduction of insects adapted as natural enemies of *C. monilifera* such as the bitou tip moth (*Comostolopsis germana*) and bitou seed fly (*Mesoclanis polana*). In Australia, while these have had some success in controlling bitou bush (ssp. *rotundata*), to date they have not had similar success in combating boneseed (ssp. *monilifera*).

In a study carried out by researchers at the University of New England and published in 2017, it has been found that a serious error was made with the introduction of biological control agents for *C. monilifera* ssp. *rotundata*. Bitou seed fly (*Mesoclanis polana*) was irresponsibly introduced into Australia based on the naive belief that it is a natural enemy of *C. monilifera*. After reviewing many hours of video footage of the bitou bush flowers in Northern NSW, researchers at the School of Ecosystem Management found that *Mesoclanis polana* is actually the most frequent pollinator of *C. monilifera*. As *C. monilifera* is a weed of National Significance in Australia, this oversight could potentially be devastating to Australian ecosystems. Much like the introduction of the cane toad to control the population of cane beetles, such a discovery is an important reminder on the importance of properly researching biological controls before introducing them into new ecosystems.

HERACLEUM MANTEGAZZIANUM

Heracleum mantegazzianum, commonly known as giant hogweed, is a monocarpic perennial herbaceous flowering plant in the carrot family Apiaceae. *H. mantegazzianum* is also known as cartwheel-flower, giant cow parsley, giant cow parsnip, or hogsbane. In New Zealand, it is also sometimes called wild parsnip (not to be confused with *Pastinaca sativa*) or wild rhubarb.

Giant hogweed is native to the western Caucasus region of Eurasia. It was introduced to Britain as an ornamental plant in the 19th century, and has also spread to other areas in Western Europe, the United States, and Canada. Its close relatives, Sosnowsky's hogweed and Persian hogweed, have similarly spread to other parts of Europe.

The sap of giant hogweed is phototoxic and causes phytophotodermatitis in humans, resulting in blisters and scars. These serious reactions are due to the furanocoumarin derivatives in the leaves, roots, stems, flowers, and seeds of the plant. Consequently, it is considered to be a noxious weed in many jurisdictions.

Green, red-spotted stem with white hairs.

Giant hogweed typically grows to heights of 2 to 5 m (6 ft 7 in to 16 ft 5 in). Under ideal conditions, a plant can reach a height of 5.5 m (18 ft). The leaves are incised and deeply lobed. A mature plant has huge leaves, between 1–1.5 m (3 ft 3 in–4 ft 11 in) wide, and a stout, bright green stem with extensive dark reddish-purple splotches and prominent coarse white hairs, especially at the base of the leaf stalk. Hollow, ridged stems vary from 3–8 cm (1.2–3.1 in) in diameter, occasionally up to 10 cm (3.9 in) in diameter and can grow to more than 4 m (13 ft) high. Dark red spots on the stem each surround a single hair. The umbrella-shaped inflorescence, called a compound umbel, may be up to 100 cm (39 in) in diameter across its flat top. The flowers are white or greenish white and may be radially symmetrical or strongly bilaterally symmetrical (zygomorphic). The fruits are schizocarps, producing seeds in dry, flattened, oval pairs. Each seed is approximately 1 cm (0.39 in) in length, with a broadly rounded base and broad marginal ridges, tan in color with brown lines (so-called oil tubes) extending 3/4 of the length of the seed.

Life Cycle

The life cycle of giant hogweed consists of four phases:

1. Pre-flowering plants: In the first year, leaves sprout from seed. In subsequent years, leaves sprout from overwintering roots as well as seeds. This pre-flowering phase continues for several years.

2. Flowering plants (midsummer): After several years of growth, the plant flowers.

3. Seeds (late summer/early autumn): A flowering plant produces 20,000 or more seeds.

4. Dead stems (late autumn/winter): After producing seeds, the plant dies, leaving dried stems and seed heads standing.

During the first few years of growth, the leaves and stem of a pre-flowering plant die over the winter. In the spring, the plant grows back from its root. In other words, the giant hogweed is a herbaceous perennial.

A giant hogweed plant usually produces a flowering stalk in 3–5 years, but plants may take up to 8 years to flower if conditions are unfavorable. In the Czech Republic, a single plant reached 12 years old before flowering. In any case, when the plant finally flowers, it does so between June and July.

Seeds are typically produced in August. A single flowering plant will produce 20,000 seeds on average with seed production varying between 10,000 and 50,000 seeds per plant.

Giant hogweed is a monocarpic perennial, that is, after a mature plant flowers and produces seed, the entire plant dies. During the following winter, tall dead stems mark the locations where the flowering plants once stood.

The seeds are dispersed short distances by wind, but can travel longer distances by water, animals, and people. The vast majority of seeds (95%) are found in the top 5 cm (2.0 in) of the soil within a few meters of the parent plant. Seeds may stay alive in the seed bank for more than five years.

A seed deposited in the seed bank is initially dormant. Dormancy is broken by the cold and wet conditions of fall and winter, and so freshly deposited seeds lay dormant until at least the following spring, at which time approximately 90% of the previously dormant seeds will germinate. The rest remain dormant in the seed bank.

Seeds normally result from cross-pollination between two or more plants but self-pollination is also possible. More than half the seeds produced by self-pollination will germinate and give rise to healthy seedlings. Hence a single isolated seed may give rise to a colony of new plants.

Similar Species

The various species of the genus *Heracleum* are similar in appearance, but vary in size. *H. mantegazzianum* is among the tallest, typically reaching 4 m (13 ft) high (and sometimes more than 5 m (16 ft) high), whereas *Heracleum* species native to Western Europe or North America, such as the cow parsnip (*H. maximum*), rarely exceed 3 m (9.8 ft) high. There are considerable differences in the size of the umbel, leaves, and stem of *H. mantegazzianum* as well.

The following table compares *Heracleum mantegazzianum* and *Heracleum maximum* feature by feature:

	H. mantegazzianum	*H. maximum*
Height	Typically 3 to 4.5 m (9.8 to 14.8 ft) tall	Up to 2.5 m (8 ft 2 in) tall
Leaves	Compound, lobed leaves typically 100 cm (39 in) wide, up to 150 cm (59 in) wide; mature leaf has deep incisions and serrated edges	Compound, lobed leaves up to 60 cm (24 in); mature leaf is less incised with less jagged edges
Stem	Green stems from 3–8 cm (1.2–3.1 in) in diameter, occasionally up to 10 cm (3.9 in) in diameter, with dark reddish-purple splotches and coarse white hairs at the base of the leaf stalk	Green ridged stems up to 5 cm (2.0 in) in diameter with fine white hairs (no purple splotches)
Flowers	White umbel is typically 80 cm (31 in) in diameter, up to 100 cm (39 in) in diameter, with 50–150 flower rays per umbel; flowers bloom mid-June to mid-July	White umbel up to 30 cm (12 in) in diameter with 15–30 flower rays per umbel; flowers bloom late May to late June
Fruits	Oval-shaped fruits Mericarps of the giant hogweed (*H. mantegazzianum*)	Heart-shaped fruits Mericarps of the common cow parsnip (*H. maximum*)

Other plant species in the family Apiaceae have features similar to those of the giant hogweed (*H. mantegazzianum*):

- In Ontario, giant hogweed look-alikes include: cow parsnip (*Heracleum maximum*), wild parsnip (*Pastinaca sativa*), various angelica species (especially *Angelica atropurpurea*, the native purplestem angelica), and Queen Anne's lace (*Daucus carota*).

- In New York State, giant hogweed look-alikes include: cow parsnip (*Heracleum maximum*), wild parsnip (*Pastinaca sativa*), angelica, and poison hemlock.

- In Europe, giant hogweed look-alikes include: hogweed (*Heracleum sphondylium* and *Heracleum sphondylium ssp sibiricum*), wild parsnip (*Pastinaca sativa*), garden angelica (*Angelica archangelica*), and wild angelica (*Angelica sylvestris*).

In Europe, over 20 species of the genus *Heracleum* have been recorded. Identification of *Heracleum mantegazzianum* is further complicated by the presence of two additional tall invasive hogweed species: *Heracleum sosnowskyi* and *Heracleum persicum*. Other than size, these three species have very similar characteristics.

Distribution

Distribution of giant hogweed in Europe.

Giant hogweed is widespread throughout western and northern Europe, especially along many terrains, such as coastal areas and riverbanks. By forming dense stands, it can displace native plants and reduce wildlife habitats. It has spread in the northeastern and northwestern United States, and southern Canada and is an invasive species across western Europe; in sites where it has settled, it overtakes the local native species, *Heracleum sphondylium.*

In Canada, the plant occurs in most provinces, except the in the prairies. It has been seen in Quebec since the early 1990s. The plant's spread in Ontario began in the southwest and was seen in 2010 in the Greater Toronto Area and Renfrew County near Ottawa.

In the United States, giant hogweed occurs in Wisconsin and south to Indiana, Michigan, Maryland, and New Jersey. In June 2018, it was reported growing in Virginia and North Carolina. The plant is a federally listed as a noxious weed in many US states.

Public Health and Safety

Giant hogweed flower head.

The sap of the giant hogweed plant is phototoxic. Contact with the plant sap prevents the skin from being able to protect itself from sunlight, which leads to phytophotodermatitis, a serious skin inflammation. A phototoxic reaction can begin as soon as 15 minutes after contact with the sap. Photosensitivity peaks between 30 minutes and two hours after contact but can last for several days. Authorities advise that all humans (especially children) should stay away from giant hogweed. Protective clothing, including eye protection, should be worn when handling the plant. Parts of the body that come into contact with the sap of giant hogweed should be immediately washed with soap and cold water, and further exposure to sunlight should be avoided for at least 48 hours. Other *Heracleum* species, such as the *cow parsnip* (*Heracleum maximum*), are likewise phototoxic and hence, similar caution is advised. Due to physical similarities to Queen Anne's lace, giant hogweed and its relatives are sometimes mistaken as harmless plants.

Control Measures

Because of its phototoxicity and invasive nature, giant hogweed is often actively removed. The European Union funded the *Giant Alien* project to combat the plant. On August 2, 2017, it added the species to its *List of Invasive Alien Species of Union concern*, thus placing restrictions on keeping, importing, selling, breeding and growing it and requiring governments to detect and eradicate it throughout the EU. In the UK, the Wildlife and Countryside Act 1981 makes it an offence to plant or cause giant hogweed to grow in the wild.

In the US, hogweed is regulated as a federal noxious weed by the US government, and is illegal to import into the United States or move interstate without a permit from the Department of Agriculture. The USDA Forest Service states pigs and cattle can eat it without apparent harm. The New York State Department of Environmental Conservation has had an active program to control giant hogweed since 2008. In 2011, Maine state horticulturists, describing the plant as "Queen Anne's lace on steroids", reported that it has been found at 21 different locations in Maine, with the number of plants ranging from one to a hundred.

IMPATIENS GLANDULIFERA

Impatiens glandulifera is a large annual plant native to the Himalayas. Via human introduction it is now present across much of the Northern Hemisphere and is considered an invasive species in many areas. Uprooting or cutting the plants is an effective means of control.

Leaves of the Himalayan Balsam.

Glands.

It typically grows to 1 to 2 m (3.3 to 6.6 ft) high, with a soft green or red-tinged stem, and lanceolate leaves 5 to 23 cm (2.0 to 9.1 in) long. The crushed foliage has a strong musty smell. Below the leaf stems the plant has glands that produce a sticky, sweet-smelling, and edible nectar. The flowers are pink, with a hooded shape, 3 to 4 cm (1 $\frac{1}{4}$ to 1 $\frac{1}{2}$ in) tall and 2 cm ($\frac{3}{4}$ in) broad; the flower shape has been compared to a policeman's helmet.

After flowering between June and October, the plant forms seed pods 2 to 3 cm ($\frac{3}{4}$ to 1 $\frac{1}{4}$ in) long and 8 mm broad ($\frac{1}{4}$ in), which explode when disturbed, scattering the seeds up to 7 metres (23 feet).

The green seed pods, seeds, young leaves and shoots are all edible. The flowers can be turned into a jam or parfait.

The plant was rated in first place for per day nectar production per flower in a UK plants survey conducted by the AgriLand project which is supported by the UK Insect Pollinators Initiative. (However, when number of flowers per floral unit, flower abundance, and phenology were taken into account it dropped out of the top 10 for most nectar per unit cover per year, as did all plants that placed in the top ten along with this one for per day nectar production per flower, with the exception of Common Comfrey, *Symphytum officinale.*)

Distribution

Himalayan Balsam is native to the Himalayas, specifically to the areas between Kashmir and Uttarakhand. In its native range it is usually found in altitudes between 2000-2500 m above sea level, although it has been reported in up to 4000 m above sea level.

In Europe the plant was first introduced in the United Kingdom where it has become naturalized and widespread across riverbanks. Presently it can be found almost everywhere across the continent.

In North America it has been found in the Canadian provinces of British Columbia, Manitoba, Ontario, Quebec, Nova Scotia, New Brunswick, Prince Edward Island and Newfoundland. In the United States it is found on both the east and west coast, seemingly restricted to northern latitudes.

In New Zealand it is sometimes found growing wild along riverbanks and wetlands.

Invasive Species

Himalayan Balsam is sometimes cultivated for its flowers. It is now widely established in other parts of the world (such as the British Isles and North America), in some cases becoming an invasive species weed. The aggressive seed dispersal, coupled with high nectar production which attracts pollinators, often allows the Himalayan Balsam to outcompete native plants. Himalayan Balsam also promotes river bank erosion due to the plant dying back over winter, leaving the bank unprotected from flooding. Invasive Himalayan Balsam can also adversely affect indigenous species by attracting pollinators (e.g. insects) at the expense of indigenous species. It is considered a "prohibited noxious weed" under the *Alberta Weed Control Act 2010.*

In the UK the plant was first introduced in 1839 at the same time as Giant Hogweed and Japanese Knotweed. These plants were all promoted at the time as having the virtues of "herculean proportions" and "splendid invasiveness" which meant that ordinary people could buy them for the cost of a packet of seeds to rival the expensive orchids grown in the greenhouses of the rich. Within ten years, however, Himalayan balsam had escaped from the confines of cultivation and begun to spread along the river systems of England. Today it has spread across most of the UK and some local wildlife trusts organise "balsam bashing" events to help control the plant. However, a recent study concludes that in some circumstances, such efforts may cause more harm than good. Destroying riparian stands of Himalayan Balsam can open up the habitat for more aggressive invasive plants such as Japanese knotweed and aid in seed dispersal (by dropped seeds sticking to shoes). Riparian habitat is suboptimal for *I. glandulifera*, and spring or autumn flooding destroys seeds and plants. The research suggests that the optimal way to control the spread of riparian Himalayan Balsam is to decrease eutrophication, thereby permitting the better-adapted local vegetation that gets outgrown by the balsam on watercourses with high nutrient load to rebound naturally. They caution that these conclusions do probably not hold true for stands of the plant at forest edges and meadow habitats, where manual destruction is still the best approach.

Himalayan Balsam.

The Bionic Control of Invasive Weeds in Wiesbaden, Germany, is trying to establish a self-sufficient project to conserve their local biodiversity by developing several food products made from the Impatiens flowers. Eventually, if all goes well, this project will have the Himalayan Balsam financing its own eradication.

In August 2014, CABI released a rust fungus in Berkshire, Cornwall and Middlesex in the United Kingdom as part of field trials into the biological control of Himalayan Balsam.

Some research also suggests that *I. glandulifera* may exhibit allelopathy, in which it excretes toxins that negatively affect neighboring plants, thus increasing its competitive advantage.

The Royal Horticultural Society and the Centre for Ecology and Hydrology recommend that pulling and cutting is the main method of non-chemical control, and usually the most appropriate. Natural Resources Wales has used manual methods such as pulling plants and using strimmers to largely eradicate Himalayan Balsam from reaches of the River Ystwyth.

REYNOUTRIA JAPONICA

Reynoutria japonica, synonyms *Fallopia japonica* and *Polygonum cuspidatum*, is a large species of herbaceous perennial plant of the knotweed and buckwheat family Polygonaceae. It is commonly known as Asian knotweed or Japanese knotweed. It is native to East Asia in Japan, China and Korea. In North America and Europe, the species has successfully established itself in numerous habitats, and is classified as a pest and invasive species in several countries.

Japanese knotweed has hollow stems with distinct raised nodes that give it the appearance of bamboo, though it is not related. While stems may reach a maximum height of 3–4 m (9.8–13.1 ft) each growing season, it is typical to see much smaller plants in places where they sprout through cracks in the pavement or are repeatedly cut down. The leaves are broad oval with a truncated base, 7–14 cm (2.8–5.5 in) long and 5–12 cm (2.0–4.7 in) broad, with an entire margin. The flowers are small, cream or white, produced in erect racemes 6–15 cm (2.4–5.9 in) long in late summer and early autumn.

Related species include giant knotweed (*Reynoutria sachalinensis*, syns. *Fallopia sachalinensis, Polygonum sachalinense*) and Russian vine (*Fallopia baldschuanica, Polygonum baldschuanicum*).

Dead stems from previous years remain in place as new growth appears.

A hedgerow made up of roses and Japanese knotweed.

Erect inflorescence.

Names

Common names for Japanese knotweed include fleeceflower, Himalayan fleece vine, billyweed, monkeyweed, monkey fungus, elephant ears, pea shooters, donkey rhubarb, American bamboo, and Mexican bamboo, among many others, depending on country and location. In Japanese, the name is *itadori*.

In Japan, the plant is known as *itadori*. The kanji expression is from the Chinese meaning "tiger stick". One interpretation of the Japanese name is that it comes from "remove pain" (alluding to its painkilling use), though there are other etymological explanations offered.

Uses

Japanese knotweed flowers are valued by some beekeepers as an important source of nectar for honeybees, at a time of year when little else is flowering. Japanese knotweed yields a monofloral honey, usually called *bamboo honey* by northeastern U.S. beekeepers, like a mild-flavoured version of buckwheat honey (a related plant also in the Polygonaceae).

A variegated variety of Japanese knotweed, used as a landscape plant.

The young stems are edible as a spring vegetable, with a flavour similar to extremely sour rhubarb. In some locations, semi-cultivating Japanese knotweed for food has been used as a means of controlling knotweed populations that invade sensitive wetland areas and drive out the native vegetation. It is eaten in Japan as *sansai* or wild foraged vegetable.

It is used in traditional Chinese and Japanese medicine to treat various disorders through the actions of resveratrol, although there is no high-quality evidence from clinical research for any medical efficacy. Extracts of resveratrol from *R. japonica* roots are higher in content than those from stems or leaves, and have highest levels at the end of the growing season.

Ground-feeding songbirds also eat the seeds.

This antique locomotive.

Invasive Species

It is listed by the World Conservation Union as one of the world's worst invasive species.

The invasive root system and strong growth can damage concrete foundations, buildings, flood defences, roads, paving, retaining walls and architectural sites. It can also reduce the capacity of channels in flood defences to carry water.

It is a frequent colonizer of temperate riparian ecosystems, roadsides and waste places. It forms thick, dense colonies that completely crowd out any other herbaceous species and is now considered one of the worst invasive exotics in parts of the eastern United States. The success of the species has been partially attributed to its tolerance of a very wide range of soil types, pH and sa-linity. Its rhizomes can survive temperatures of −35 °C (−31 °F) and can extend 7 metres (23 ft) horizontally and 3 metres (9.8 ft) deep, making removal by excavation extremely difficult. The plant is also resilient to cutting, vigorously resprouting from the roots.

Identification

Identification of Japanese knotweed is not always easy. Many other plants are suspected of being knotweed, due often to the similar appearance of leaves and stems. Dogwood, lilac, Houttuynia (*Houttuynia cordata*), ornamental Bistorts such as Red Bistort (*Persicaria amplexicaulis*), less-er knotweed (*Koenigia campanulata*), Himalayan Balsam (*Impatiens glandulifera*), Broadleaved Dock (*Rumex obtusifolius*), Bindweed (*Convolvulus arvensis*), bamboo, Himalayan Honeysuckle (*Leycesteria formosa*), and Russian Vine (*Fallopia baldschuanica*) have been suspected of being *Reynoutria japonica*.

New leaves of *Reynoutria japonica* are dark red and 1 to 4 cm (0.4 to 1.6 in) long; young leaves are green and rolled back with dark red veins; leaves are green and shaped like a heart flattened at the base, or a shield, and are usually around 12 cm (4.7 in) long. Mature *R. japonica* forms 2-to-3-metre (6.6 to 9.8 ft) tall dense thickets; stems look somewhat like bamboo, with rings and purple speckles. Leaves shoot from the stem nodes alternately in a zigzag pattern. Mature stems are hollow and not at all woody: they can be snapped easily to see if they are hollow. Plants that are immature or affected by mowing or other restrictions have much thinner and shorter stems than mature stands, and are not hollow.

Control

Japanese knotweed has a large underground network of roots (rhizomes). To eradicate the plant the roots need to be killed. All above-ground portions of the plant need to be controlled repeatedly for several years in order to weaken and kill the entire patch. Picking the right herbicide is essen-tial, as it must travel through the plant and into the root system below.

The abundance of the plant can be significantly reduced by applying glyphosate, imazapyr, a com-bination of both, or by cutting all visible stalks and filling the stems with glyphosate. However, these methods have not been proven to provide reliable long-term results in completely eliminat-ing the treated population.

Digging up the rhizomes is a common solution where the land is to be developed, as this is quicker than the use of herbicides, but safe disposal of the plant material without spreading it is difficult; knotweed is classed as controlled waste in the UK, and disposal is regulated by law. Digging up the roots is also very labour-intensive and not always efficient. The roots can go to up to 10 feet (3.0 me-ters) deep, and leaving only a few inches of root behind will result in the plant quickly growing back.

Covering the affected patch of ground with a non-translucent material can be an effective fol-low-up strategy. However, the trimmed stems of the plant can be razor sharp and are able to pierce through most materials. Covering with non-flexible materials such as concrete slabs has to be done meticulously and without leaving even the smallest splits. The slightest opening can be enough for the plant to grow back.

More ecologically-friendly means are being tested as an alternative to chemical treatments. Soil steam sterilization involves injecting steam into contaminated soil in order to kill subterranean plant parts. Research has also been carried out on *Mycosphaerella* leafspot fungus, which devastates knotweed in its native Japan. This research has been relatively slow due to the complex life cycle of the fungus.

Two biological pest control agents that show promise in the control of the plant are a leaf spot fungus from genus *Mycosphaerella* and the psyllid *Aphalara itadori*. Following earlier studies imported Japanese knotweed psyllid insects (*Aphalara itadori*), whose only food source is Japanese knotweed, were released at a number of sites in Britain in a study running from 1 April 2010 to 31 March 2014. In 2012, results suggested that establishment and population growth were likely, after the insects overwintered successfully.

Anecdotal reports of effective control describe the use of goats to eat the plant parts above ground followed by the use of pigs to root out and eat the underground parts of the plant.

Detail of the stalk.

The most effective method of control is by herbicide application close to the flowering stage in late summer or autumn. In some cases, it is possible to eradicate Japanese knotweed in one growing season using only herbicides. Glyphosate is widely used as it is non-persistent, and certain formulations may be used in or near water.

Trials in Haida Gwaii, British Columbia, using sea water sprayed on the foliage, have not demonstrated promising results.

PILOSELLA AURANTIACA

Pilosella aurantiaca (fox-and-cubs, orange hawk bit, devil's paintbrush, grim-the-collier) is a perennial flowering plant in the daisy family Asteraceae that is native to alpine regions of central and southern Europe, where it is protected in several regions.

It is a low-growing plant with shallow fibrous roots and a basal rosette of elliptical to lanceolate leaves 5–20 centimetres (2.0–7.9 in) long and 1–3 centimetres (0.39–1.18 in) broad. All parts of the plant exude a milky juice. The flowering stem is usually leafless or with just one or two small leaves. The stem and leaves are covered with short stiff hairs (trichomes), usually blackish in color. The stems may reach a height of 60 centimetres (24 in) and have 2–25 capitula (flowerheads), each 1–2 $\frac{1}{2}$ cm diameter, bundled together at the end of short pedicels. The flowers are orange, almost red, which is virtually invisible to bees, yet they also reflect ultraviolet light, increasing their conspicuousness to pollinators. The flowers are visited by various insects, including many species of bees, butterflies, pollinating flies. The flowers themselves come in a range of colors from a deep rust-orange to a pure yellow and often show striking gradients of color.

The plant propagates through its wind-dispersed seeds, and also vegetatively by stolons and shallow rhizomes.

Cultivation and Uses

Whole plant.

P. aurantiaca is widely grown as an ornamental plant in gardens for its very decorative flowers. often used in wildflower gardens due to its bright orange flowers being highly attractive to a wide array of pollinators.

Invasive Weed

Orange hawkweed is currently the only hawkweed considered regionally invasive in areas of British Columbia, Canada. It is considered invasive in the East Kootenay, Central Kootenay, Columbia-Shuswap, Thompson-Nicola, Bulkley Nechako, and Cariboo Regional Districts. Invasive hawkweed can replace native vegetation in open, undisturbed natural areas such as meadows, reducing forage and threatening biodiversity. In Victoria and NSW, Australia, hawkweed species are declared as "State Prohibited Weeds" and are controlled under The Bio Security Act 2015. Currently there are several eradication programs operating (often employing volunteers) to locate, prevent the spread of and eradicate any *Pilosella* or *Hieracium* plants.

ECHIUM PLANTAGINEUM

Echium plantagineum, commonly known as purple viper's-bugloss or Paterson's curse, is a species of *Echium* native to western and southern Europe (from southern England south to Iberia and east to the Crimea), northern Africa, and southwestern Asia (east to Georgia). It has also been introduced to Australia, South Africa and United States, where it is an invasive weed. Due to a high concentration of pyrrolizidine alkaloids, it is poisonous to grazing livestock, especially those with simple digestive systems, like horses.

Echium plantagineum is a winter annual plant growing to 20–60 cm tall, with rough, hairy, lanceolate leaves up to 14 cm long. The flowers are purple, 15–20 mm long, with all the stamens protruding, and borne on a branched spike.

Invasive Species

Echnium plantagineum has become an invasive species in Australia, where it is also known as Salvation Jane (particularly in South Australia), blueweed, Lady Campbell weed, Paterson's Curse and Riverina bluebell.

In the United States, the species has become naturalised in parts of California, Oregon, and some eastern states and areas such as northern Michigan. In Oregon it has been declared a noxious weed.

Toxicity

Echium plantagineum contains pyrrolizidine alkaloids and is poisonous. When eaten in large quantities, it causes reduced livestock weight and death, in severe cases, due to liver damage. Paterson's curse can also kill horses, and irritate the udders of dairy cows and the skin of humans. After the 2003 Canberra bushfires, a large bloom of the plant occurred on the burned land, and many horses became ill and died from grazing on it. Because the alkaloids can also be found in the nectar of Paterson's curse, the honey made from it should be blended with other honeys to dilute the toxins.

ASPARAGUS ASPARAGOIDES

Asparagus asparagoides, commonly known as bridal creeper, bridal-veil creeper, *gnarboola*, smilax or smilax asparagus, is a herbaceous climbing plant of the family Asparagaceae native to eastern and southern Africa. Sometimes grown as an ornamental plant, it has become a serious environmental weed in Australia and New Zealand.

Asparagus asparagoides grows as a herbaceous vine with a scrambling or climbing habit which can reach 3 m (10 ft) in length. It has shiny green leaf-like structures (phylloclades) which are flattened stems rather than true leaves. They measure up to 4 cm long by 2 cm wide. The pendent white flowers appear over winter and spring, from July to September. It is rhizomatous, and bears tubers which reach 6 cm (2.4 in) by 2 cm (1.8 in) in size.

Distribution and Habitat

It ranges throughout tropical Africa, south to Namibia, and the fynbos in South Africa, as far south as Cape Town.

It has become naturalised in parts of southern California.

Uses

Asparagus asparagoides, often under the name smilax, is commonly used in floral arrangements or home decorating.

Invasive Species

A bridal creeper infestation.

A. asparagoides is a major weed species in southern Australia and in New Zealand. In Australia, it is listed as a Weed of National Significance.

It was introduced to Australia from South Africa around 1857, for use as a foliage plant, especially in bridal bouquets (hence the common name). It has escaped into the bush and smothers the native vegetation with the thick foliage and thick underground mat of tubers which restrict root growth of other species. It is recognised as one of the 20 "weeds of national significance". The seeds are readily spread in the droppings of birds, rabbits and foxes, as well as the plant extending its root system. CSIRO have introduced several biological controls in an attempt to reduce the spread and impact of the weed.

In New Zealand *A. asparagoides* is listed under the National Pest Plant Accord and is classified as an "unwanted organism".

A. asparagoides has also escaped cultivation in California.

CYNODON DACTYLON

Cynodon dactylon, also known as *Vilfa stellata*, Bermuda grass, *Dhoob, dūrvā grass, dubo*, dog's tooth grass, Bahama grass, devil's grass, couch grass, Indian *doab, arugampul, grama*, wiregrass and scutch grass, is a grass that originated in Africa. Although it is not native to Bermuda, it is an abundant invasive species there. It is presumed to have arrived in North America from Bermuda, resulting in its common name. In Bermuda it has been known as *crab grass* (also a name for *Digitaria sanguinalis*).

The blades are a grey-green colour and are short, usually 2–15 cm (0.79–5.91 in) long with rough edges. The erect stems can grow 1–30 cm (0.39–11.81 in) tall. The stems are slightly flattened, often tinged purple in colour.

The seed heads are produced in a cluster of two to six spikes together at the top of the stem, each spike 2–5 cm (0.79–1.97 in) long.

It has a deep root system; in drought situations with penetrable soil, the root system can grow to over 2 metres (6.6 ft) deep, though most of the root mass is less than 60 centimetres (24 in) under the surface. The grass creeps along the ground with its stolons and roots wherever a node touches the ground, forming a dense mat. *C. dactylon* reproduces through seeds, stolons, and rhizomes. Growth begins at temperatures above 15 °C (59 °F) with optimum growth between 24 and 37 °C (75 and 99 °F); in winter, the grass becomes dormant and turns brown. Growth is promoted by full sun and retarded by full shade, e.g., close to tree trunks.

Cultivation, Control and Uses

Cynodon dactylon is widely cultivated in warm climates all over the world between about 30° S and 30° N latitude, and that get between 625 and 1,750 mm (24.6 and 68.9 in) of rainfall a year (or less, if irrigation is available). It is also found in the U.S., mostly in the southern half of the country and in warm climates. Detailed study using experimental animals exhibited anti-stress, adaptogenic activities and improved male fertility.

Control/Eradication

It is fast-growing and tough, making it popular and useful for sports fields, as when damaged it will recover quickly. It is a highly desirable turf grass in warm temperate climates, particularly for those regions where its heat and drought tolerance enable it to survive where few other grasses do. This combination makes it a frequent choice for golf courses in the southern and southeastern U.S. It has a relatively coarse-bladed form with numerous cultivars selected for different turf requirements. It is also highly aggressive, crowding out most other grasses and invading other habitats, and has become a hard-to-eradicate weed in some areas (it can be controlled somewhat with Triclopyr, Mesotrione, Fluazifop-P-butyl, and Glyphosate). This weedy nature leads some gardeners to give it the name of "devil grass". Bermuda grass is incredibly difficult to control in flower beds and most herbicides do not work. However, Ornamec, Ornamec 170, Turflon ester (tricyclopyr), and Imazapyr have shown some effectiveness. All of these items are difficult to find in retail stores, as they are primarily marketed to professional landscapers.

Bermuda grass has been cultivated in saline soils in California's Central Valley, which are too salt-damaged to support agricultural crops; it was successfully irrigated with saline water and used to graze cattle.

The hybrid variety Tifton 85, like some other grasses (e.g. sorghum), produces cyanide under certain conditions, and has been implicated in several livestock deaths (note that in several places this variety has been incorrectly reported as a genetically modified strain; actually it is a conventionally bred F$_1$ hybrid).

In India, commonly known as "durva", this grass is used in the Ayurveda system of medicine. In Hinduism, it is considered important in the worship of Lord Ganesha.

 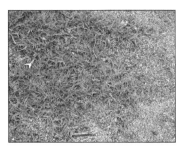

Cynodon dactylon.

Varieties

- Tifgreen (drought resistant),

- Tifway 419 or Tifton 419 (athletic fields, fairways, golf course tees),

- LaPaloma,

- Riviera,

- SR9554,

- Laprima,

- Veracruz,

- Wrangler,

- Yukon,

- AgriDark (sports turf – Australia and New Zealand),

- OZTUFF (low maintenance couch – Australia, PBR under OZ-E-GREEN),

- This list is not all inclusive. Hundreds of cultivars have been created specifically for environmental tolerance and stakeholder requirements. New cultivars are released yearly.

EUPHORBIA VIRGATA

Euphorbia virgata, commonly known as leafy spurge, wolf's milk leafy spurge, or wolf's milk is a species of spurge native to Europe and Asia, and naturalized in North America, where it is an invasive species.

It has commonly been confused with *Euphorbia esula*, a species no longer considered a permanent component of the North American flora. *Euphorbia esula* is restricted to certain parts of Europe and not considered a weedy species, while *E. virgata* is found throughout the United States and Canada and has caused significant economic and ecological impacts. *E. virgata* is best distinguished from *E. esula* by its leaves, which are 6-15 times longer than wide with margins that are (near-)parallel at the middle of the blade, while *E. esula* leaves are wider toward the tip, usually 3-8 times longer than wide, with margins that are not parallel at the middle of the leaf. In addition, the apex of *E. virgata* is usually acute and the base is truncate to attenuate, while the apex of *E. esula* is rounded or subacute and the base is more gradually attenuate or cuneate. Mentions of *E. esula* in the North American flora and invasive species literature are now referred to *E. virgata*.

As an Invasive Plant

Leafy spurge was transported to the United States possibly as a seed impurity in the early 19th century. It now occurs across much of the northern U.S., with the most extensive infestations reported for Montana, North Dakota, Nebraska, South Dakota, and Wyoming. Since the introduction of leafy spurge to the United States, it has been found in 458 counties in 26 states. It has been identified as a serious weed on a number of national parks and on reserves of The Nature Conservancy in eleven northern states. It is now classified as an invasive species by the United States Department of Agriculture.

Leafy spurge grows in a variety of different climate environments. It displaces native vegetation in prairie habitats and fields through shading and by usurping available water and nutrients and through plant toxins that prevent the growth of other plants underneath it. It is an aggressive invader and, once present, can completely overtake large areas of open land. Leafy spurge quickly colonizes in areas with bare soil, especially those caused by human disturbance where native species are removed. One method of control suggested is to limit the amount of bare soil from these disturbances. It is toxic as well.

Leafy spurge quarantine sign.

Because of its persistent nature and ability to regenerate from small pieces of root, leafy spurge is extremely difficult to eradicate. In Saskatchewan, leafy spurge spread on trails used by settlers and traders. More recently, the travel of leafy spurge has been linked to railroads and the use and transport of infested hay used for agriculture. Biological control offers a highly promising management tactic for leafy spurge. Goats, apparently able to graze on the plant without ill effect, have been used on rail trails in Idaho to clear leafy spurge from the trail shoulders. Sheep have been used in North Dakota, along with herbicides and flea beetles, to fight it. The U.S. Department of

Agriculture has shown success using six European insects that feed on leafy spurge. These include a stem and root-boring beetle (*Oberea erythrocephala*), five root-mining flea beetles (*Aphthona* spp.), The Spurge Hawk-moth (*Hyles euphorbiae*), and a shoot-tip gall midge (*Spurgia esulae*). Large scale field-rearing and release programs are carried out cooperatively by federal and state officials in many northern U.S. states. The results are not as immediate as when herbicides are used but, if pesticide use is kept to a minimum, large numbers of these insects build up within a few years and have shown impressive results.

Several systemic herbicides have been found to be effective if applied in June, when the flowers and seeds are developing, or in early-to-mid-September, when the plants are moving nutrients downward into the roots. Preliminary research suggests that chemical treatment in the fall followed by a spring burn to reduce seed germination may be an effective strategy for reducing leafy spurge infestations. Multiple treatments are necessary every year for several years, making leafy spurge control an extremely expensive undertaking. If left uncontrolled for a single year, leafy spurge can reinfest rapidly. Prescribed burning, in conjunction with herbicides, may also be effective.

As a Model Weed

Leafy spurge is being developed as a model to answer fundamental questions of weed biology. Over 55,000 expressed sequence tags have been sequenced from all plant tissues including tissues from plants that were cold stressed, drought stressed, or attacked by both flea beetles and gall midges. Analysis of the EST sequences indicated that 23,000 unique sequences representing more than 19,000 unigenes were obtained. These sequences are now available on Genbank. The unigenes have been used to develop cDNA microarrays that also include more than 4,000 additional cDNAs from cassava (another Euphorb related to leafy spurge). These microarrays are being used to identify physiological processes and signals that regulate bud dormancy (one of the main reasons leafy spurge is difficult to control) and invasiveness.

AILANTHUS ALTISSIMA

Ailanthus altissima commonly known as tree of heaven, ailanthus, varnish tree, or in Chinese as *chouchun*, is a deciduous tree in the family Simaroubaceae. It is native to both northeast and central China, as well as Taiwan. Unlike other members of the genus *Ailanthus*, it is found in temperate climates rather than the tropics. The tree grows rapidly and is capable of reaching heights of 15 metres (49 ft) in 25 years. While the species rarely live more than 50 years, some specimens exceed 100 years of age. Its remarkable suckering ability makes it possible for this tree to clone itself almost indefinitely. It is considered a noxious weed and vigorous invasive species.

In China, the tree of heaven has a long and rich history. It was mentioned in the oldest extant Chinese dictionary and listed in many Chinese medical texts for its purported curative ability. The roots, leaves and bark are used in traditional Chinese medicine, primarily as an astringent. The tree has been grown extensively both in China and abroad as a host plant for the ailanthus silkmoth, a moth involved in silk production. Ailanthus has become a part of western culture as well, with the tree serving as the central metaphor and subject matter of the best-selling American novel *A Tree Grows in Brooklyn* by Betty Smith.

The tree was first brought from China to Europe in the 1740s and to the United States in 1784. It was one of the first trees brought west during a time when *chinoiserie* was dominating European arts, and was initially hailed as a beautiful garden specimen. However, enthusiasm soon waned after gardeners became familiar with its suckering habits and its foul smelling odor. Despite this, it was used extensively as a street tree during much of the 19th century. Outside Europe and the United States the plant has been spread to many other areas beyond its native range. In a number of these, it has become an invasive species due to its ability both to colonize disturbed areas quickly and to suppress competition with allelopathic chemicals. It is considered a noxious weed in Australia, the United States, New Zealand and many countries of central, eastern and southern Europe. The tree also resprouts vigorously when cut, making its eradication difficult and time-consuming. This has led to the tree being called "tree of hell" among gardeners and conservationists.

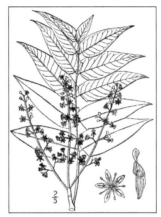
Botanical drawing of the leaves, flowers and samaras.

A. altissima is a medium-sized tree that reaches heights between 17 and 27 metres (56 and 89 ft) with a diameter at breast height of about 1 metre (39 inches). The bark is smooth and light grey, often becoming somewhat rougher with light tan fissures as the tree ages. The twigs are stout, smooth to lightly pubescent, and reddish or chestnut in color. They have lenticels as well as heart-shaped leaf scars (i.e. a scar left on the twig after a leaf falls) with many bundle scars (i.e. small marks where the veins of the leaf once connected to the tree) around the edges. The buds are finely pubescent, dome shaped, and partially hidden behind the petiole, though they are completely visible in the dormant season at the sinuses of the leaf scars. The branches are light to dark gray in color, smooth, lustrous, and containing raised lenticels that become fissures with age. The ends of the branches become pendulous. All parts of the plant have a distinguishing strong odor that is often likened to peanuts, cashews, or rotting cashews.

The leaves are large, odd- or even-pinnately compound, and arranged alternately on the stem. They range in size from 30 to 90 cm (0.98 to 2.95 ft) in length and contain 10–41 leaflets organised in pairs, with the largest leaves found on vigorous young sprouts. When they emerge in the spring, the leaves are bronze then quickly turn from medium to dark green as they grow. The rachis is light to reddish-green with a swollen base. The leaflets are ovate-lanceolate with entire margins, somewhat asymmetric and occasionally not directly opposite to each other. Each leaflet is 5 to 18 cm (2.0 to 7.1 in) long and 2.5 to 5 cm (0.98 to 1.97 in) wide. They have a long tapering end while the bases have two to four teeth, each containing one or more glands at the tip. The leaflets' upper sides are dark green in color with light green veins, while the undersides are a more whitish green.

The petioles are 5 to 12 mm (0.20 to 0.47 in) long. The lobed bases and glands distinguish it from similar sumac species.

Bark and flowers of *A. altissima*.

Immature seeds on a female tree.

Tree of Heaven Re-sprouting even after herbicide use to restore Red Butte Creek.

The flowers are small and appear in large panicles up to 50 cm (20 in) in length at the end of new shoots. The individual flowers are yellowish green to reddish in color, each with five petals and sepals. The sepals are cup-shaped, lobed and united while the petals are valvate (i.e. they meet at the edges without overlapping), white and hairy towards the inside. They appear from mid-April in the south of its range to July in the north. *A. altissima* is dioecious, with male and female flowers being borne on different individuals. Male trees produce three to four times as many flowers as the females, making the male flowers more conspicuous. Furthermore, the male plants emit a foul-smelling odor while flowering to attract pollinating insects. Female flowers contain ten (or rarely five through abortion) sterile stamens (stamenoides) with heart-shaped anthers. The pistil is made up of five free carpels (i.e. they are not fused), each containing a single ovule. Their styles are united and slender with star-shaped stigmas. The male flowers are similar in appearance, but they of course lack a pistil and the stamens do function, each being topped with a globular anther and a glandular green disc. The fruits grow in clusters; a fruit cluster may contain hundreds of seeds. The seeds borne on the female trees are 5 mm in diameter and each is encapsulated in a samara that is 2.5 cm (0.98 in) long and 1 cm (0.39 in) broad, appearing July though August, but can persist on the tree until the next spring. The samara is large and twisted at the tips, making it spin as it falls, assisting wind dispersal, and aiding buoyancy for long-distance dispersal through hydrochory. Primary wind dispersal and secondary water dispersal are usually positively correlated in *A. altissima* since most morphological characteristics of samaras affect both dispersal modes in the same way – except for the width of the samaras, which in contrast affects both types of dispersal in opposing ways, allowing differentiation in the dispersal strategies of this tree. The females can produce huge amounts of seeds, normally around 30,000 per kilogram (14,000/lb) of tree, and fecundity can be estimated non-destructively through measurements of dbh.

Taxonomy

The first scientific descriptions of the tree of heaven were made shortly after it was introduced to Europe by the French Jesuit Pierre Nicholas d'Incarville. D'Incarville had sent seeds from Peking via Siberia to his botanist friend Bernard de Jussieu in the 1740s. The seeds sent by d'Incarville were thought to be from the economically important and similar looking Chinese varnish tree (*Toxicodendron vernicifluum*), which he had observed in the lower Yangtze region, rather than the tree of heaven. D'Incarville attached a note indicating this, which caused much taxonomic confusion over the next few decades. In 1751, Jussieu planted a few seeds in France and sent others on to Philip Miller, the superintendent at the Chelsea Physic Garden, and to Philip C. Webb, the owner of an exotic plant garden in Busbridge, England.

Confusion in naming began when the tree was described by all three men with three different names. In Paris, Linnaeus gave the plant the name *Rhus succedanea*, while it was known commonly as *grand vernis du Japon*. In London the specimens were named by Miller as *Toxicodendron altissima* and in Busbridge it was dubbed in the old classification system as *Rhus Sinese foliis alatis*. There are extant records from the 1750s of disputes over the proper name between Philip Miller and John Ellis, curator of Webb's garden in Busbridge. Rather than the issue being resolved, more names soon appeared for the plant: Jakob Friedrich Ehrhart observed a specimen in Utrecht in 1782 and named it *Rhus cacodendron*.

Light was shed on the taxonomic status of ailanthus in 1788 when René Louiche Desfontaines observed the samaras of the Paris specimens, which were still labelled *Rhus succedanea*, and came to the conclusion that the plant was not a sumac. He published an article with an illustrated description and gave it the name *Ailanthus glandulosa*, placing it in the same genus as the tropical species then known as *A. integrifolia* (white siris, now *A. triphysa*). The name is derived from the Ambonese word *ailanto*, meaning "heaven-tree" or "tree reaching for the sky". The specific *glandulosa*, referring to the glands on the leaves, persisted until as late as 1957, but it was ultimately made invalid as a later homonym at the species level. The current species name comes from Walter T. Swingle who was employed by the United States Department of Plant Industry. He decided to transfer Miller's older specific name into the genus of Desfontaines, resulting in the accepted name *Ailanthus altissima*.

There are three varieties of *A. altissima*:

- *A. altissima* var. *altissima*, which is the type variety and is native to mainland China.

- *A. altissima* var. *tanakai*, which is endemic to northern Taiwan highlands. It differs from the type in having yellowish bark, odd-pinnate leaves that are also shorter on average at 45 to 60 cm (18 to 24 in) long with only 13–25 scythe-like leaflets. It is listed as endangered in the IUCN Red List of threatened species due to loss of habitat for building and industrial plantations.

- *A. altissima* var. *sutchuenensis*, which differs in having red branchlets.

Distribution and Habitat

A. altissima is native to northern and central China, Taiwan and northern Korea. It was historically widely distributed, and the fossil record indicates clearly that it was present in North America as recently as the middle Miocene. In Taiwan it is present as var. *takanai*. In China it is native to every province except Gansu, Heilongjiang, Hainan, Jilin, Ningxia, Qinghai, Xinjiang, and Tibet. It has been introduced in many regions across the world.

The tree prefers moist and loamy soils, but is adaptable to a very wide range of soil conditions and pH values. It is drought-hardy, but not tolerant of flooding. It also does not tolerate deep shade. In China it is often found in limestone-rich areas. The tree of heaven is found within a wide range of climatic conditions. In its native range it is found at high altitudes in Taiwan as well as lower ones in mainland China. These are virtually found anywhere in the U.S., but especially in arid regions bordering the Great Plains, very wet regions in the southern Appalachians, cold areas of the lower Rocky Mountains and throughout much of the California Central Valley, forming dense thickets that displace native plants. Prolonged cold and snow cover cause dieback, although the trees resprout from the roots.

Ailanthus altissima.

As an Exotic Plant

The earliest introductions of *A. altissima* to countries outside of its native range were to the southern areas of Korea as well as to Japan. It is possible that the tree is native to these areas, but it is generally agreed that the tree was a very early introduction. Within China itself it has also been naturalized beyond its native range in areas such as Qinghai, Ningxia and Xinjiang.

In 1784, not long after Jussieu had sent seeds to England, some were forwarded to the United States by William Hamilton, a gardener in Philadelphia. In both Europe and America it quickly became a favoured ornamental, especially as a street tree, and by 1840 it was available in most nurseries. The tree was separately brought to California in the 1890s by Chinese immigrants who came during the California Gold Rush. It has escaped cultivation in all areas where it was introduced, but most extensively in the United States. It has naturalized across much of Europe, including Germany, Austria, Switzerland, the Czech Republic, the Pannonian region (i.e. southeastern Central Europe around the Danube river basin from Austria, Slovakia and Hungary south to the Balkan ranges) and most countries of the Mediterranean Basin. In Montenegro and Albania *A. altissima* is widespread in both rural and urban areas and while in the first it was introduced as an ornamental plant, it very soon invaded native ecosystems with disastrous results and became an invasive species. Ailanthus has also been introduced to Argentina, Australia (where it is a declared weed in New South Wales and Victoria), New Zealand (where it is listed under the National Pest Plant Accord and is classed an "unwanted organism"), the Middle East and in some countries in South Asia such as Pakistan. In South Africa it is listed as an invasive species which must be controlled, or removed and destroyed.

In North America, *A. altissima* is present from Massachusetts in the east, west to southern Ontario, southwest to Iowa, south to Texas, and east to the north of Florida. On the west coast it is found from New Mexico west to California and north to Washington. In the east of its range it grows most extensively in disturbed areas of cities, where it was long ago present as a planted street tree. It also grows along roads and railways. For example, a 2003 study in North Carolina found the tree of heaven was present on 1.7% of all highway and railroad edges in the state and had been expanding its range at the rate of 4.76% counties per year. Similarly, another study conducted in southwestern Virginia determined that the tree of heaven is thriving along approximately 30% of the state's interstate highway system length or mileage. It sometimes enters undisturbed areas as well and competes with native plants. In western North America it is most common in mountainous areas around old dwellings and abandoned mining operations. It is classified as a noxious or invasive plant on National Forest System lands and in many states because its prolific seed production, high germination rates and capacity to regrow from roots and root fragments enable *A. altissima* to out-compete native species. For this reason, control measures on public lands and private property are advised where *A. altissima* has naturalized.

Ecology

A female bearing a heavy load of seeds.

Tree of heaven is an opportunistic plant that thrives in full sun and disturbed areas. It spreads aggressively both by seeds and vegetatively by root sprouts, re-sprouting rapidly after being cut. It is considered a shade-intolerant tree and cannot compete in low-light situations, though it is sometimes found competing with hardwoods, but such competition rather indicates it was present at the time the stand was established. On the other hand, a study in an old-growth hemlock-hardwood forest in New York found that Ailanthus was capable of competing successfully with native trees in canopy gaps where only 2 to 15% of full sun was available. The same study characterised the tree as using a "gap-obligate" strategy in order to reach the forest canopy, meaning it grows rapidly during a very short period rather than growing slowly over a long period. It is a short lived tree in any location and rarely lives more than 50 years. Ailanthus is among the most pollution-tolerant of tree species, including sulfur dioxide, which it absorbs in its leaves. It can withstand cement dust and fumes from coal tar operations, as well as resist ozone exposure relatively well. Furthermore, high concentrations of mercury have been found built up in tissues of the plant.

Ailanthus has been used to re-vegetate areas where acid mine drainage has occurred and it has been shown to tolerate pH levels as low as 4.1 (approximately that of tomato juice). It can withstand very low phosphorus levels and high salinity levels. The drought-tolerance of the tree is strong due to its ability to effectively store water in its root system. It is frequently found in areas where few trees can survive. The roots are also aggressive enough to cause damage to subterranean sewers and pipes. Along highways it often forms dense thickets in which few other tree species are present, largely due to the toxins it produces to prevent competition.

Female tree.

Ailanthus produces an allelopathic chemical called ailanthone, which inhibits the growth of other plants. The inhibitors are strongest in the bark and roots, but are also present in the leaves, wood and seeds of the plant. One study showed that a crude extract of the root bark inhibited 50% of a sample of garden cress (*Lepidium sativum*) seeds from germinating. The same study tested the extract as an herbicide on garden cress, redroot pigweed (*Amaranthus retroflexus*), velvetleaf (*Abutilon theophrasti*), yellow bristlegrass (*Setaria pumila*), barnyard grass (*Echinochloa crusgalli*), pea (*Pisum sativum* cv. Sugar Snap) and maize (*Zea mays* cv. Silver Queen). It proved able to kill nearly 100% of seedlings with the exception of velvetleaf, which showed some resistance. Another experiment showed a water extract of the chemical was either lethal or highly damaging to 11 North American hardwoods and 34 conifers, with the white ash (*Fraxinus americana*) being the only plant not adversely affected. The chemical does not, however, affect the tree of heaven's own seedlings, indicating that *A. altissima* has a defence mechanism to prevent autotoxicity. Resistance in various plant species has been shown to increase with exposure. Populations without prior exposure to the chemicals are most susceptible to them. Seeds produced from exposed plants have also been shown to be more resistant than their unexposed counterparts.

Leaves in autumn.

The tree of heaven is a very rapidly growing tree, possibly the fastest growing tree in North America. Growth of one to two metres (3.3 to 6.6 ft) per year for the first four years is considered normal. Shade considerably hampers growth rates. Older trees, while growing much slower, still do so faster than other trees. Studies found that Californian trees grew faster than their East Coast counterparts, and American trees in general grew faster than Chinese ones.

In northern Europe the tree of heaven was not considered naturalised in cities until after the Second World War. This has been attributed to the tree's ability to colonize areas of rubble of destroyed buildings where most other plants would not grow. In addition, the warmer microclimate in cities offers a more suitable habitat than the surrounding rural areas (it is thought that the tree requires a mean annual temperature of 8 degrees Celsius to grow well, which limits its spread to more northern and higher altitude areas). For example, one study in Germany found the tree of heaven growing in 92% of densely populated areas of Berlin, 25% of its suburbs and only 3% of areas outside the city altogether. In other areas of Europe this is not the case as climates are mild enough for the tree to flourish. It has colonized natural areas in Hungary, for example, and is considered a threat to biodiversity at that country's Aggtelek National Park.

Several species of Lepidoptera utilise the leaves of ailanthus as food, including the Indian moon moth (*Actias selene*) and the common grass yellow (*Eurema hecabe*). In North America the tree is the host plant for the ailanthus webworm (*Atteva aurea*), though this ermine moth is native to Central and South America and originally used other members of the mostly tropical Simaroubaceae as its hosts. In its native range *A. altissima* is associated with at least 32 species of arthropods and 13 species of fungi.

In North America, the leaves of ailanthus are sometimes attacked by *Aculops ailanthii*, a mite in the family Eriophyidae. Leaves infested by the mite begin to curl and become glossy, reducing their ability to function. Therefore, this species has been proposed as a possible biocontrol for ailanthus in the Americas.

Due to the tree of heaven's weedy habit, landowners and other organisations often resort to various methods of control in order to keep its populations in check. For example, the city of Basel in Switzerland has an eradication program for the tree. It can be very difficult to eradicate, however. Means of eradication can be physical, thermal, managerial, biological or chemical. A combination of several of these can be most effective, though they must of course be compatible. All have some positive and negative aspects, but the most effective regimen is generally a mixture of chemical and physical control. It involves the application of foliar or basal herbicides in order to kill existing trees, while either hand pulling or mowing seedlings in order to prevent new growth.

Uses

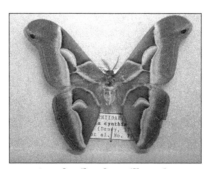

A male ailanthus silkmoth.

In addition to its use as an ornamental plant, the tree of heaven is also used for its wood and as a host plant to feed silkworms of the moth *Samia cynthia*, which produces silk that is stronger and cheaper than mulberry silk, although with inferior gloss and texture. It is also unable to take dye. This type of silk is known under various names: "pongee", "eri silk" and "Shantung silk", the last name being derived from Shandong Province in China where this silk is often produced. Its production is particularly well known in the Yantai region of that province. The moth has also been introduced in the United States.

The pale yellow, close-grained and satiny wood of ailanthus has been used in cabinet work. It is flexible and well suited to the manufacture of kitchen steamers, which are important in Chinese cuisine for cooking mantou, pastries and rice. Zhejiang Province in eastern China is most famous for producing these steamers. It is also considered a good source of firewood across much of its range as it is moderately hard and heavy, yet readily available. The wood is also used to make charcoal for culinary purposes. There are problems with using the wood as lumber, however. Because the trees exhibit rapid growth for the first few years, the trunk has uneven texture between the inner and outer wood, which can cause the wood to twist or crack during drying. Techniques have been developed for drying the wood so as to prevent this cracking, allowing it to be commercially harvested. Although the live tree tends to have very flexible wood, the wood is quite hard once properly dried.

Cultivation

Tree of heaven is a popular ornamental tree in China and valued for its tolerance of difficult growing conditions. It was once very popular in cultivation in both Europe and North America, but this popularity dropped, especially in the United States, due to the disagreeable odor of its blossoms and the weediness of its habit. The problem of odor was previously avoided by only selling pistillate plants since only males produce the smell, but a higher seed production also results. Michael Dirr, a noted American horticulturalist and professor at the University of Georgia, reported meeting, in 1982, a grower who could not find any buyers. He further writes:

> "For most landscaping conditions, it has *no* value as there are too many trees of superior quality; for impossible conditions this tree has a place; selection could be made for good habit, strong wood and better foliage which would make the tree more satisfactory; I once talked with an architect who tried to buy *Ailanthus* for use along polluted highways but could not find an adequate supply."

In Europe, however, the tree is still used in the garden to some degree as its habit is generally not as invasive as it is in America. In the United Kingdom it is especially common in London squares, streets, and parks, though it is also frequently found in gardens of southern England and East Anglia. It becomes rare in the north, occurring only infrequently in southern Scotland. It is also rare in Ireland. In Germany the tree is commonly planted in gardens. The tree has furthermore become unpopular in cultivation in the west because it is short-lived and that the trunk soon becomes hollow, making trees more than two feet in diameter unstable in high winds.

A few cultivars exist, but they are not often sold outside of China and probably not at all in North America:

- 'Hongye' – The name is Chinese and means "red leaves". As the name implies it has attractive vivid red foliage.

- 'Thousand Leaders'

- 'Metro' – A male cultivar with a tighter crown than usual and a less weedy habit.

- 'Erythrocarpa' – The fruits are a striking red.

- 'Pendulifolia' – Leaves are much longer and hang elegantly.

Traditional Medicine

Nearly every part of *A. altissima* has some application in Chinese traditional medicine. One of the oldest recipes, recorded in a work from AD 732, is used for treating mental illness. It involved chopped root material, young boys' urine and douchi. After sitting for a day, the liquid was strained out and given to an ill person over the course of several days.

Another source from 684 AD, during the Tang dynasty and recorded in Li Shizhen's *Compendium of Materia Medica*, states that when the leaves are taken internally, they make one incoherent and sleepy, while when used externally, they can be effectively used to treat boils, abscesses and itches. This formula calls for young leaves of ailanthus, catalpa and peach tree to be crushed together and the resulting liquid applied to the scalp to stimulate hair growth.

The dried bark, however, is listed in the modern Chinese materia medica as *chun bai pi*, meaning "white bark of spring". Modern works treat it in detail, discussing chemical constituents, how to identify the product and its possible uses. It is prepared by felling the tree in fall or spring, stripping the bark and then scraping off the hardest, outermost portion, which is then sun-dried, soaked in water, partially re-dried in a basket and finally cut into strips. The bark is said to have cooling and astringent properties. It is only prescribed in amounts between 4 and 10 grams, so as not to poison people. Li's Compendium has 18 recipes that call for the bark. A tincture of the root-bark has been claimed to have been used successfully by American herbalists in the 19th century. It contains phytochemicals, such as quassin and saponin, and ailanthone. It is available in most shops dealing in Chinese traditional medicine. The samaras are also used in modern Chinese medicine under the name *feng yan cao*, meaning "herbal phoenix eye".

The plant may be mildly toxic. The noxious odours have been associated with nausea and headaches, as well as with contact dermatitis reported in both humans and sheep, which developed weakness and paralysis. It contains a quinone irritant, 2,6-dimethoxybenzoquinone, as well as quassinoids.

CYPERUS ESCULENTUS

Cyperus esculentus (also called chufa sedge, nut grass, yellow nutsedge, tiger nutsedge, edible galingale, water grass or earth almond) is a crop of the sedge family widespread across much of the world. It is found in most of the Eastern Hemisphere, including Southern Europe, Africa and Madagascar, as well as the Middle East and the Indian subcontinent. *C. esculentus* is cultivated for its edible tubers, called earth almonds or tiger nuts, as a snackfood and for the preparation of *horchata de chufa*, a sweet, milk-like beverage.

Cyperus esculentus can be found wild, as a weed, or as a crop. It is an invasive species in its native range, and is readily transported accidentally to become invasive. In many countries, *C. esculentus* is considered a weed. It is often found in wet soils such as rice paddies and peanut farms as well as well irrigated lawns and golf courses during warm weather.

Botany

Cyperus esculentus is an annual or perennial plant, growing to 90 cm (3.0 ft) tall, with solitary stems growing from a tuber. The plant is reproduced by seeds, creeping rhizomes, and tubers. Due to its clonal nature, *C. esculentus* can take advantage of soil disturbances caused by anthropogenic or natural forces. The stems are triangular in section and bear slender leaves 3–10 mm (1/8 to 1/2 inches) wide. The spikelets of the plant are distinctive, with a cluster of flat, oval seeds surrounded by four hanging, leaf-like bracts positioned 90 degrees from each other. They are 5 to 30 mm (about 3/8 to 1 1/8 inches) long and linear to narrowly elliptic with pointed tips and 8 to 35 florets. The color varies from straw-colored to gold-brown. They can produce up to 2420 seeds per plant. The plant foliage is very tough and fibrous and is often mistaken for a grass. The roots are an extensive and complex system of fine, fibrous roots and scaly rhizomes with small, hard, spherical tubers and basal bulbs attached.

The tubers are 0.3 – 1.9 cm (1/8 to 3/4 inches) in diameter and the colors vary between yellow, brown, and black. One plant can produce several hundred to several thousand tubers during a single growing season. With cool temperatures, the foliage, roots, rhizomes, and basal bulbs die, but the tubers survive and resprout the following spring when soil temperatures remain above 6 °C (43 °F). They can resprout up to several years later. When the tubers germinate, many rhizomes are initiated and end in a basal bulb near the soil surface. These basal bulbs initiate the stems and leaves above ground, and fibrous roots underground. *C. esculentus* is wind pollinated and requires cross pollination as it is self–incompatible.

Invasiveness

C. esculentis is a highly invasive species in Oceania, Mexico, some United States, and the Caribbean, mainly by seed dispersion. It is readily transported internationally, and is adaptable to re-establish in varied climate and soil environments. In Japan, it is an exotic clonal weed favorable to establish in wet habitats.

Cultivation

Cultivation and growing of the *xufa*.

Climate Requirements

Cyperus esculentus cultivation requires a mild climate. Low temperature, shade, and light intensity can inhibit flowering. Tuber initiation is inhibited by high levels of nitrogen, long photoperiods, and high levels of gibberellic acid. Flower initiation occurs under photoperiods of 12 to 14 hours per day.

Soil Requirements

Tubers can develop in soil depths around 30 cm (1-foot), but most occur in the top or upper part. They tolerate many adverse soil conditions including periods of drought and flooding and survive soil temperatures around –5 °C (23 °F). They grow best on sandy, moist soils at a pH between 5.0 – 7.5. The densest populations of *C. esculentus* are often found in low-lying wetlands. They do not tolerate salinity.

Agronomy

Cultivation

Planting is normally done on flat soils where ridges to favour the coming irrigations have previously been done. The separation between ridges is approximately 60 cm (2.0 ft) and seeds are planted manually. Distances between seeds may vary from 15 to 20 cm (6 to 8 in) and seeding depth is around 8 cm (3 in). A typical seeding rate for chufa is about 120 kg of tubers/ha (107 lbs/acre).

They are planted between April and May and must be irrigated every week until they are harvested in November and December. Tubers develop about 6 – 8 weeks after seedling emergence and grow quickly during July and August. The maturing is around 90 – 110 days. The average yield can approach between 10 and 19 t/ha.

Harvest and Drying Process

Harvest usually occurs in November or December and the leaves are scorched during the harvest. With a combine harvester, the tiger nut is pulled out of the ground. Immediately after harvesting, the tiger nuts are washed with water in order to remove sand and small stones. The drying occurs usually in the sun and can take up to three months. The temperatures and humidity levels have to be monitored very carefully during this period. The tiger nuts have to be turned every day to ensure uniform drying. The drying process ensures a longer shelf life. This prevents rot or other bacterial infections, securing quality and high nutrition levels. Disadvantages in the drying process are shrinkage, skin wrinkles and hard nut texture.

Storage

Tiger nut loses a considerable amount of water during drying and storage. The starch content of the tiger nut tubers decreases and the reducing sugar (invert sugar) content increases during storage. Tiger nut can be stored dry and rehydrated by soaking without losing the crisp texture. Soaking is often done overnight. Dried tiger nuts have a hard texture and soaking is indispensable to render them edible with ease and to ensure acceptable sensory quality.

According to the *Consejo Regulador de Chufa de Valencia* (Regulating Council for Valencia's Tiger Nuts), the nutritional composition/100 ml of the Spanish beverage horchata de chufas is as follows: energy content around 66 kcal, proteins around 0.5 g, carbohydrates over 10 g with starch at least 1.9 g, fats at least 2 g.

Uses

Dried tiger nut has a smooth, tender, sweet, and nutty taste. It can be consumed raw, roasted, dried, baked or as tiger nut milk or oil.

Food

Dried tubers sold at the market.

The tubers are edible, with a slightly sweet, nutty flavour, compared to the more bitter-tasting tuber of the related *Cyperus rotundus* (purple nutsedge). They are quite hard and are generally soaked in water before they can be eaten, making them much softer and giving them a better texture. They are a popular snack in West Africa.

They have various uses, such as horchata, a nonalcoholic beverage of milky appearance derived from the tubers of the tiger nut plant mixed with sugar and water, and commonly consumed in Spain.

Flour of roasted tiger nut is sometimes added to biscuits and other bakery products as well as in making oil, soap, and starch extracts. It is also used for the production of nougat, jam, beer, and as a flavoring agent in ice cream and in the preparation of kunnu (a local beverage in Nigeria). Kunnu is a nonalcoholic beverage prepared mainly from cereals (such as millet or sorghum) by heating and mixing with spices (dandelion, alligator pepper, ginger, licorice) and sugar. To make up for the poor nutritional value of kunnu prepared from cereals, tiger nut was found to be a good substitute for cereal grains. Tiger nut oil can be used naturally with salads or for deep frying. It is considered to be a high quality oil. Tiger nut "milk" has been tried as an alternative source of milk in fermented products, such as yogurt production, and other fermented products common in some African countries and can thus be useful replacing milk in the diet of people intolerant to lactose to a certain extent.

Nutrition

Despite its name, tiger nutsedge is a tuber. However, its chemical composition shares characteristics with tubers and with nuts. This tuber is rich in energy content (starch, fat, sugar, and protein), and dietary minerals (mainly phosphorus and potassium). The oil of the tuber was found to contain 18% saturated (palmitic acid and stearic acid) and 82% unsaturated (oleic acid and linoleic acid) fatty acids.

Oil

Since the tubers of *C. esculentus* contain 20-36% oil, it has been suggested as potential oil crop for the production of biodiesel. One study found that chufa produced 1.5 metric tonnes of oil per hectare (174 gallons/acre) based on a tuber yield of 5.67 t/ha and an oil content of 26.4%. A similar 6-year study found tuber yields ranging from 4.02 to 6.75 t/ha, with an average oil content of 26.5% and an average oil yield of 1.47 t/ha.

Fishing Bait

The boiled nuts are used in the UK as a bait for carp. The nuts have to be prepared in a prescribed manner to prevent harm to the fish. The nuts are soaked in water for 24 hours, and then boiled for 20 minutes or longer until fully expanded. Some anglers then leave the boiled nuts to ferment for 24–48 hours, which can enhance their effectiveness. If the nuts are not properly prepared, they can be toxic to carp. This was originally thought to have been the cause of death of Benson, a large, well-known female carp weighing 54 lb (24 kg) found floating dead in a fishing lake, with a bag of unprepared tiger nuts lying nearby, empty, on the bank. An examination of the fish by a taxidermist concluded tiger nut poisoning was not the cause of death, but rather the fish had died naturally.

Compatibility with other Crops

The seed head of a *Cyperus esculentus* plant.

C.esculentus is extremely difficult to remove completely once established. This is due to the plant having a stratified and layered root system, with tubers and roots being interconnected to a depth of 36 cm or more. The tubers are connected by fragile roots that are prone to snapping when pulled, making the root system difficult to remove intact. Intermediate rhizomes can potentially reach a length of 60 cm. The plant can quickly regenerate if a single tuber is left in place. By competing for light, water and nutrients it can reduce the vigour of neighbouring plants. It can develop into a dense colony. Patch boundaries can increase by more than one meter per year. Tubers and seed disperse with agricultural activities, soil movement or by water and wind. They are often known as a contaminant in crop seeds. When plants are small they are hard to distinguish from other weeds such as Dactylis glomerata and Elytrigia repens. Thus it is hard to discover in an early stage and therefore hard to counteract. Once it is detected, mechanical removal, hand removal, grazing, damping, and herbicides can be used to inhibit *C.esculentus*.

Similar native or non-native species that can confuse identification:

- Sedges (*Cyperus*) have grass-like leaves and resemble each other in the appearance. They can mainly be distinguished from grasses by their triangular stems.

- Purple nutsedge (*C. rotundus*) is another weedy sedge that is similar to the yellow nutsedge (*C. esculentus*). These two sedges are difficult to distinguish from each other and can be found growing on the same site. Some differences are the purple spikelets and the tubers formed by *C.rotondus* are often multiple instead of just one at the tip. In addition the tubers have a bitter taste instead of the mild almond-like flavour of *C.esculentus*.

PURPLE LOOSESTRIFE

Purple loosestrife is native to Europe and Asia. It was introduced to North America in the early 1800s in ship ballast and as a medicinal herb. It is now found in 40 US states.

- A semi-aquatic perennial species that typically forms a dense bushy growth of many erect stems reaching heights of approximately 4- 7 feet tall. It is highly visible from July through September because of its robust purple flowering spikes.

- Leaves are smooth-edged, slender, pointed and arranged in opposite pairs along ridged stems.

- Showy spikes of flowers develop at the tops of each stem consisting of many individual 5- 7 petaled purple flowers.

- Large roots develop over time and store high levels of nutrients providing the plant with reserves of energy early in the spring or during stressful periods.

Bee feeding on flower.

Flower.

Plants.

Leaves.

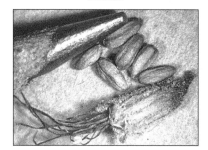

Mimosa pudica with mature seed pods on plant.

Infestation.

Plant.

Infestation in wetland.

Habitat

Purple loosestrife prefers wet soils or standing water. Loosestrife plants are typically found in poorly drained soils of road right-of-ways and trails, drainage ditches, culverts, lake shores, stream banks, and a variety of wetland habitats.

Means of Spread and Distribution

Purple loosestrife reproduces both by seed and vegetative propagation which allows it to quickly invade new landscapes. Each flower spike can produce thousands of tiny seeds that are easily dispersed by wind, water, snow, animals, and humans.

Impact

Purple loosestrife aggressively invades lakes, rivers, and wetlands, creates large monocultures, and significantly decreases the biological diversity of native plant and wildlife populations.

Prevention and Management

- A sound management plan will take several years of commitment, especially on older stands that have an established seed bank. Regular follow-up is critical to ensure the population is decreasing.

- Hand pulling or digging is only recommended when a few plants are discovered on a property. To successfully control purple loosestrife in this manner, the entire root system has to be removed from the soil to prevent re-sprouting of new stems. Checking the site periodically for several years is recommended to ensure that new seedlings or re-sprouts can be destroyed.

- Mowing or cutting is not practical for sites where loosestrife is growing in an aquatic or semi-aquatic environment. However, if conditions permit, and if executed prior to flowering, mowing or cutting can reduce seed production. Re-sprouts will vigorously appear following mowing, so follow-up cutting will be necessary to prevent seed production during the growing season. Make sure to wash equipment thoroughly following mowing to prevent spread of seeds to new areas.

- Various herbicides have been used successfully against purple loosestrife. Due to the fact that purple loosestrife is a semi-aquatic to aquatic species, it is important to use only herbicides that are labeled and approved for use in or around water. If treating plants near water with herbicide, please be aware of the state pesticide laws and use only products labeled for aquatic use.

- Biological control, using host-specific natural enemies of purple loosestrife, is a popular form of management for this species. Biological control agents feed specifically on purple loosestrife plants and have been shown to provide a long-term sustainable management solution.

Uses of Weeds

Weeds are beneficial and have a number of applications in the environment. It includes fertilizing the soil, increasing moisture, acting as shelter, repelling pests, attracting beneficial insects, serving as food to humans, etc. All these diverse uses of weeds have been carefully analyzed in this chapter.

WEEDS AS HUMAN FOOD

Weeds abound in urban and agricultural environments. Depending on region and site, up to 66% of weed species are edible, and may constitute an additional food source for humans.

The nutritional value of the plants depends on the vegetation period. For some weeds, flowers are harvested when they are still in buds and for others, the best time is immediately after development. Depending on the species, you can cut the whole inflorescence, pluck individual flowers or just tear the petals.

Leaves are tastiest and healthiest when they are young and tender although they are not suitable for long-term storage. Roots should be harvested when the plant is dormant, in the spring or autumn.

Wild Amaranth

Also known as "Pigweed", amaranth leaves are treated as a green leafy vegetable like spinach. The seeds of wild amaranth are edible too and can be roasted. They are a good source of free protein.

The young leaves of pigweed are soft and mild in taste and can also be used in salads or teas and the older leaves can be cooked like spinach. It contains proteins, vitamin A and C and minerals.

This plant can be toxic to livestock animals due to the presence of nitrates in the leaves.

Queen Anne's Lace

The wild carrot is almost identical to the highly poisonous hemlock. There are many ways to identify wild carrot, but one important identifier of wild carrot is the smell, it smells like a carrot.

Like carrots, its roots are also edible when young (first year) but can become woody if not harvested on time. Its flower heads are edible too and can be eaten raw or cooked.

Lamb's Quarters

Lamb's quarter leaves tend to look dusty and have a white powdery coating on them. This edible weed is packed with nutrition. Its tender leaves are great in salads and can be used as a substitute for other leafy vegetables. Its flowers and seeds are also edible that tastes like quinoa. However, its seeds contain toxic called Saponins in meager amount and should not be consumed in excess. Lamb's quarters contain some oxalic acid therefore when eating this raw, small quantities are recommended.

However, its seeds contain toxic called Saponins in meager amount and should not be consumed in excess.

This plant can be toxic to livestock animals due to the presence of nitrates in the leaves.

Comfrey

Comfrey is a member of the borage family. Although not very aromatic, this herb is known for it medicinal properties and high protein content. Comfrey roots and leaves are also used to treat wounds because they contain allantoin, a substance that helps new skin cells grow to heal the wound quickly.

Comfrey can also be used as a garden fertilizer and as mulch.

Winter Cress

The Winter Cress is available at the time in winter when most the plants don't even grow. It belongs to the mustard family and considered as a weed. However, it is a rich source of Vitamin C.

The leaves are bitter but best in taste before the plant starts to flower while they are still young and tender. At this stage, they can be added to salads like rocket. It can also be used as a vegetable like spinach.

Common Mallow

Common mallow (*Malva sylvestris*) has many medicinal and edible uses and can be easy found growing wild in most places. All parts of this plant are edible. The leaves, flowers, fruits and seeds can be eaten, whether raw in salads or cooked and like many other leafy greens, usually more tender and tasty when they are smaller and less mature. More mature leaves can be cooked.

It has a very mild flavor, although the plant is quite rich in vitamins A, B, and C, along with calcium, magnesium, and potassium.

Purslane

Also called "Pussley", the common Purslane (*Portulaca oleracea*) is a healthy edible weed from the moss rose family. Also, this nutritious succulent has more omega-3 fatty acids than any other leafy edible plant according to researchers at the University of Texas at San Antonio. It can be a great addition to a salad, soups or stews. It has a crunchy texture and leaves and stems can be eaten raw or cooked to add a spicy flavor to any dish.

Chickweed

Chickweed grows in a unique, intertwined manner, and it has small white star-shaped flowers. The stems have a thin line of white hair that grows in a weave-like pattern. Chickweed's stems, leaves, and flowers are all edible. This delicious weed has a mild, refreshing flavor. The leaves and stems can be added to salads like lettuces or cooked as greens. It is loaded with nutritious elements and has many medicinal uses too.

Chickweed's stems, leaves, and flowers are all edible. This delicious weed has a mild, refreshing flavor. The leaves and stems can be added to salads like lettuces or cooked as greens. It is loaded with nutritious elements and has many medicinal uses too.

Plantain

The plantain is found in meadows, pastures, roadsides or in neglected sites. The herbaceous plant is up to 50 centimeters high, the narrow leaves are up to 25 centimeters long. It erupts a mushroom-like aroma. Plantain can be used in a preparation of soups, salads or as a vegetable. The juice of crushed leaves helps against itching.

Nettle

The stinging nettle is one of the most easily identifiable weed and invasive too. Nettle is also rich in vitamin A and C, iron, magnesium, calcium and antioxidants. The young leaves are cooked. Its soup is also popular in many countries.

Dandelion

Dandelion has a very bad reputation as a weed, especially among those who like to keep a clean and green lawn. However, this plant is pollinator's favorite and also this plant is edible, from the roots to the flowers. The dandelion leaves can be harvested too at any time in the growing season.

Common Burdock

This biennial weed is very common along the ditches but also in the mountains at low altitude. The first-year roots and second-year stems can be cooked by boiling for about 20 minutes, then season to taste. Before cooking, the stems must be peeled and roots scrubbed in order to remove the bitter rind. Immature flower stalks may also be harvested in late spring before flowers appear; their taste is similar to that of artichoke. The Japanese have been known to eat the leaves when a plant is young and leaves are soft.

Clover

Beyond occasional hunting, this common herb goes unnoticed on the lawn but it is very important for bees and bumblebees. The flowers and leaves of clover can be used to add variety to meals. A few raw clover leaves can be chopped in salads or sauteed and added to dishes for a green accent; and flowers, whether red or white, can be eaten raw or cooked, including dried for tea. The refreshing acid flavor spices salads or vegetables.

Watercress

Watercress is rich in minerals, beta-carotene, and vitamin C. It is rich in nutrients. This perennial plant has hollow stems and small heart-shaped leaves. The peppery flavor of watercress is sharp but not bitter. This water-loving plant can be found growing near creeks and ponds or nearby to other waterbodies. There are many delicious watercress recipes you can easily find on the web.

Shepherd's Purse

This plant is related to mustard family, it looks like dandelion and tastes like rocket. The other distinguishable feature of Shepherd's purse is that it is up to 50 centimeters high and have white flowers. You can find this plant on sunny, nitrogen-rich clay, sand or gravel soils. The taste is somewhat reminiscent of rocket and its leaves can be eaten raw or cooked. The leaves are a watercress and cabbage substitute and become peppery with age. The flowering shoots can be eaten as well.

Water Spinach

In most tropical parts it is considered as a weed. This semi-aquatic plant is a creeper that grows in or near the water on moist soils and has light green ovate leaves, its stems are hollow, so they can float on the water. Both the leaves and stems are edible and can be used as the way you use spinach. Under tropical conditions, water spinach can be harvested throughout the year, as the leaves grow again after harvesting. In cooler climates it is grown as annual, mostly in wide containers.

MEDICINAL USES OF WEEDS

In herbal medicine, what we call weeds are commonly used as tonics, relaxants and painkillers, for digestive complaints, as antiseptics and bactericides, headaches colds and flu, and for skin disorders.

Some short-lived species rely heavily on toxic chemical defences to deter herbivores – these plants are usually avoided by grazing animals unless drought conditions or overgrazing occur. These compounds accumulate on leaves, shoots, flowers and fruits. They include glycosides, alkaloids, and terpenoids, which are all low molecular weight, often toxic at small doses, and highly biologically active. As a result, many of these types of weeds are used in the treatment of a wide variety of diseases; their effects include diuretic, choleretic, anti-inflammatory, antioxidative, anti-carcinogenic, analgesic, anti-hyperglycemic, anti-coagulatory and pre-biotic effects.

As with pharmaceutical drugs, herbal medicines can have serious side effects. In general, it's always a good idea to do a little research and consult your health-care provider for more serious conditions.

Herbal remedies are prepared in several standardized ways which usually vary based upon the plant utilized, and sometimes, what condition is being treated. Methods include: infusions (hot teas), decoctions (boiled teas), tinctures (alcohol and water extracts), and macerations (cold-soaking), and the making of salves and poultices. For some conditions, steam inhalation may be used.

As for edible weeds, there are so many medicinal 'weeds' that we cannot list them here; we will use the same species as for food to illustrate the widespread health benefits:

Dandelion (Taraxacum Officinale)

Roots of Dandelion are used to treat liver complaints, aid digestion, and can relieve food allergies. The antioxidants in Dandelions are believed to have cancer fighting power. They have also been used to repair damage that may have been caused by drugs, chemicals, alcohol, and infections conditions, like hepatitis.

Purslane (Portulaca Oleracea)

Purslane (or Pigweed) is also valued as a liver tonic. It is extremely high in omega-3 fatty acids which strengthen cell membranes – this improves the immune system, provides protection from cancers and regulates metabolism: Correcting blood pressure, cholesterol and triglycerides. It contains dopamine which has been found to be advantageous for strengthening the pituitary gland and treating Parkinson's Disease. Purslane is highly alkaline and is helpful in alleviating acidic stomachs. It is also an excellent source of vitamins A and C, iron, calcium, potassium and magnesium.

Common Chickweed (Stellaria Media)

Chickweed leaves eaten raw or cooked, are rich in iron, and a good source of calcium, chromium, cobalt, molybdenum, magnesium, manganese, silicon, zinc, vitamins A and C and antioxidants. Used as poultice or ointment. Common Chickweed in an ancient remedy for burns, rashes, bites and other skin conditions, as well as for haemorrhoids, painful joints, tendons and ligaments. It can be used fresh or dried, as a powder, ointment, or decoction. It can relieve digestive conditions like inflammation, ulceration, and bowel disorders. Respiratory complaints also benefit from Chickweed. The leaves are also eaten for arthritis, rheumatism, blood poisoning, constipation, colitis, gastritis, acid indigestion, diabetes, candida, cancer, fatigue, fractures, mouth ulcers, to strengthen the heart, improve eyesight and to assist the function of the thyroid, liver, gall, kidneys, bladder and lymphatic system.

Slender Celery (Cyclospermum Leptophyllum)

'Ajamoda' is an important drug in Indian Medicine (Ayurveda, Siddha and Unani systems) made from the fruits of some members of the carrot family. Slender Celery seeds are used in the formulations of the drug though it frequently contains seeds of related umbelliferous plants viz. Apium graveolens (Celery), and Trachispermum roxburgianum. It improves digestion and is also used to relieve bronchitis, asthma, hiccough, carminative, stimulant, cardial and pain in the bladder.

Plantain (Plantago spp.)

Plantain seeds have a high mucilage content. Soaked in hot water they will form a thick, jelly. Drunk in a fruit juice, this can help to stimulate the reflex action of the bowel. This mucin lines the mucous membranes of the digestive tract, protecting from irritation, acidity and inflammation. This mucin also benefits the mucous membranes of the throat, lungs, kidneys and urinary tubes.

Plantain leaves are anti-bacterial and anti-inflammatory and are used on cuts and abrasions. Allantoin in the leaves helps knit the cells back together and reduce inflammation. Use the leaves as a wash, as a poultice and, internally, as a tea or edible leafy greens.

There is a long list of other therapeutic uses: to correct high cholesterol and blood pressure, to boost the immune system, for candida, thrush, diarrhoea, digestive conditions and gastrointestinal ulcers, constipation, cancer, haemorrhoids; glandular complaints, spleen, bladder, kidney, liver and lung disorders; skin conditions, diabetes, irritable bowel syndrome, and as a tonic for the blood. For insect bites, crush a plantain leaf in the hands and rub it on the bite.

Weed Ecology and Biology

Weed ecology studies the interaction between weeds and its environment. The study of the growth, reproduction and life cycles of weed species and its vegetation is called weed biology. This chapter has been carefully written to provide an easy understanding of weed ecology and biology.

PROPAGATION AND DISSEMINATION OF WEED SEEDS

Propagation is the process of multiplying or increasing the number of plants of the same species and at the same time perpetuating their desirable characteristics. Plants may be propagated under two general categories: sexual and asexual propagation.

Reproduction by Seed

Reproduction by seed is called sexual reproduction. It requires pollination and fertilization of an egg which results in seed that is capable of producing a new plant. Seed production varies greatly among and within weed species in part due to environmental variability between years, competition from neighboring plants, and genetic variability. For example, while Canada thistle has been observed to produce as few as 680 seeds per plant, Curly dock often produces more than 30,000 seeds per plant.

Canada Thistle

Vegetative Reproduction

In vegetative (asexual) reproduction, a new plant develops from a vegetative organ such as a stem, root, or leaf. Several modifications of these organs are common in perennial weeds, such as underground stems (rhizomes), above-ground stems (stolons), bulbs, corms, and tubers. Although vegetative structures generally do not survive as long in the soil as do seeds, very small structures can result in a new plant. Canada thistle, for example, can produce a new plant from as small as a 1/4-inch section of root.

Vegetative reproduction can be as prolific as seed production. Yellow nut-sedge (*Cyperus esculentus*) has been reported to produce more than 1,900 new plants and more than 6,800 tubers in 1 year.

Cyperus esculentus - tubers.

Dissemination/Dispersal

A plant seed is a unique genetic entity, a biological individual. However, a seed is in a diapause state, an essentially dormant condition, awaiting the ecological conditions that will allow it to grow into an adult plant, and produce its own seeds. Seeds must therefore germinate in a safe place, and then establish themselves as a young seedling, develop into a juvenile plant, and finally become a sexually mature adult that can pass its genetic material on to the next generation.

The chances of a seed developing are generally enhanced if there is a mechanism for dispersing to an appropriate habitat some distance from the parent plant. The reason for dispersal is that closely related organisms have similar ecological requirements. Obviously, competition with the parent plant will be greatly reduced if its seeds have a mechanism to disperse some distance away. Their ability to spread and remain viable in the soil for years makes eradication nearly impossible.

Seeds have no way to move on their own, but they are excellent travelers. Plants have evolved various mechanisms that disperse their seeds effectively. Many species of plants have seeds with anatomical structures that make them very buoyant, so they can be dispersed over great distances by the winds. In the absence of proper means of their dispersal, weeds could not have moved from one country to another. An effective dispersal of weed seeds and fruits requires two essentials a successful dispersing agent and an effective adaptation to the new environment.

There are two Ways of looking at Weed Seed Dispersal

1. The expanding range and increasing population size of an invading weed species into a new area.

2. The part of the process by which an established and stabilized weed species in an area maintains itself within that area.

Dissemenation includes two separate processes. They are Dispersal (leaving mother plant) and Post-dispersal events (subsequent movement). Dispersal of seed occurs in 4 dimensions.

1. Length,

2. Width: Land/habitat/soil surface area phenomena,

3. Height (soil depth, in the air),

4. Time: shatters immediately after ripening (or) need harvesting activity to release seed Common weed dispersal agents are Wind, Water, Animals, Human, Machinery, etc.

a) Wind

Many seeds are well adapted to wind travel. Cottony coverings and parachute-like structures allow seeds to float with the wind. Examples of wind-dispersed seeds include common milkweed (*Asclepias syriaca*), common dandelion, Canada thistle, and perennial sowthistle (*Sonchus arvensis*). Weed seeds and fruits that disseminate through wind possess special organs to keep them afloat. Such organs are:

Pappus – It is a parachute like modification of persistent calyx into hairs.

Eg. Asteraceae family weeds - *Tridax procumbens*.

Tridax procumbens.

Pappus.

1. Comose: Some weed seeds are covered with hairs, partially or fully Eg. *Calotropis sp.*

2. Feathery, persistent styles - Styles are persistent and feathery Eg. *Anemone sp.*

3. Baloon - Modified papery calyx that encloses the fruits loosely along with entrapped air. Eg. *Physalis minima.*

4. Wings - One or more appendages that act as wings. Eg. *Acer macrophyllum.*

Factors that Influence Wind Dispersal

- Seed weight,

- Seed shape,

- Structures (wings or pappus),

- Height of release,

- Wind speed and turbulence.

Water

Aquatic weeds disperse largely through water. They may drift either as whole plants, plant fragments or as seeds with the water currents. Terrestrial weed seeds also disperse through irrigation and drainage water.Weed seed often moves with surface water runoff into irrigation water and ponds, where it is carried to other fields. Weeds growing in ditch banks along irrigation canals and ponds are the major source of weed seed contamination of irrigation water.

Weed seed often remains viable in water for several years, creating a "floating seedbank" and allowing weeds to disperse over large areas in moving water. Field bindweed seed, for example, remains over 50 percent viable after being submerged in water for more than 4 years. Some seeds have special adaptations that aid in water travel. The seedpod of curly dock, for example, is equipped with pontoons that carry the floating seed.

Curly Dock

Plant

Seed

Mature seed

Animals

Several weed species produce seeds with barbs, hooks, spines, and rasps that cling to the fur of animals or to clothing and then can be dispersed long distances. Farm animals carry weed seeds and fruits on their skin, hair and hooves. This is aided by special appendages such as Hooks (*Xanthium strumarium*), Stiff hairs (*Cenchrus spp*), Sharp spines (*Tribulus terrestris*) and Scarious bracts (*Achyranthus aspera*). Even ants carry a huge number of weed seeds. Donkeys eat Prosophis julifera pods.

Xanthium strumarium – hook.

Cenchrus spp - stiff hair.

Tribulus terrestris - spines.

Weed seed often is ingested and passed through the digestive tracts of animals. Animal droppings provide an ideal nutrient and moisture environment for weed germination. While only a small percentage of the seed remains viable after exposure to an animal's digestive enzymes. The ingested weed seeds are passed in viable form with animal excreta (0.2% in chicks, 9.6% in calves, 8.7% in horses and 6.4% in sheep), which is dropped wherever the animal moves. This mechanism of weed dispersal in called endozoochory. Eg., Lantana seeds by birds. Loranthus seeds stick on beaks of birds. Viable weed seeds are present in the dung of farm animals, which forms part of the FYM. Besides, addition of mature weeds to compost pit as farm waste also act as source.

Dispersal by Man

Man disperses numerous weed seeds and fruits with raw agricultural produce. Weeds mature at the same time and height along with crop, due to their similar size and shape as that of crop seed man unknowingly harvest the weeds also, and aids in dispersal of weed seeds. Such weeds are called "Satellite weeds" Eg. *Avena fatua, Phalaris minor.*

Avena fatua.

Phalaris minor.

Dispersal by Machinery

Weed seeds often are dispersed by tillage and harvesting equipment. Seeds move from field to field on the soil that sticks to tractor tires, and vegetative structures often travel on tillage and cultivation equipment and latter dropping them in other fields to start new infestation. Disctype cultivation equipment is less likely to drag vegetative plant parts than are shovels or sweeps.

Intercontinental Movement of Weeds

Introduction of weeds from one continent to another through crop seed, feed stock, packing material and nursery stock. Eg. *Parthenium hysterophorus.*

Crop Mimicry Dispersal

Weed seed adaptations to look like crop seed: plant body or seed same size, shape, and morphology as crop. Eg: Barnyard grass biotype looking like rice escapes hand weeding and is dispersed with rice, nightshade fruit ("berries") same size, shape as dry beans, harvested and dispersed with beans.

Add Mixtures with Crop Seed, Animal Feed, Hay and Straw

Weeds probably are spread more commonly during the seeding of a new crop or in animal feed and bedding than by any other method. Seed labels often indicate a tiny percentage of weed seed, but consider this example. If a legume seed contains 0.001 percent dodder (a parasitic annual; *Cuscuta campestris*) seed by weight, there will be eight dodder seeds per 2 kg of legume seed. If the legume seed is sown in a field despite an extremely low dodder seed percentage by weight, the small size of the seed, combined with rapid early-season growth, could result in an infested legume field within a single season.

LIFE CYCLE OF WEEDS

Annual Weeds

An annual is a plant which germinates, flowers, sets seed, and dies within a single year. All annuals spread only by seed. There are two types of annuals:

- Summer annuals - Summer annual weeds generally emerge as soon as soil temperatures warm in the spring or early summer. Many species continue to germinate throughout the summer under adequate conditions. Summer annual weeds grow, flower, produce seed, and are killed by frost during the fall season. Summer annual weeds are often difficult to manage, as many species are better suited to summer conditions than desirable cool-season turfgrass species. Examples are crabgrass, knotweed, and prostrate spurge.

- Winter annuals - Winter annuals germinate from seed in the late summer or early fall. Young winter annual plants live through the winter then flower, set seed and die out the following summer. Winter annuals generally cannot survive the hot summer months. Occasionally, winter annuals will germinate in the spring, but even spring-germinating weeds die out the following summer. Some examples of winter annuals are shepherd's purse, common chickweed, yellow rocket, and annual bluegrass.

Biennial Weeds

Biennial weeds usually live for two years. In the first year seeds germinate and grow without flowering, forming what is called a rosette. A rosette is a plant form with no central stalk. All leaves in a rosette arise from close to the soil surface, as in thistle.

In the second year of growth, biennials send up a flowering stalk. After flowering and seed production, biennials die. Many familiar weeds and wildflowers are biennials. Queen Anne's lace (wild carrot), evening primrose, burdock, common mullein, and moth mullein are examples of biennials found in our area. Biennials can have large taproots, which sometimes causes people to confuse them with perennials.

Perennial Weeds

A perennial is a plant which lives for many years, and does not die after flowering. All perennials have underground parts that store food over the winter and allow them to reemerge in the spring. One way to tell if a weed is a perennial is to dig it up and look for these underground parts. There are two different types of perennials. The two types are very different in their importance as weeds.

- Simple perennials - Simple perennials are also called solitary perennials. This is because these plants grow singly. Even though you may sometimes see several plants close to each other, the plants have separate root systems and are not joined underground.

Simple perennials spread only by seed. A plant which grows from a seedling may live for many years, getting larger through the years. Most simple perennials have taproots - large roots that grow vertically down through the soil. Taproots can grow to be quite large. Curly dock, plantains, and dandelion are some familiar simple perennials.

- Spreading perennials - Spreading perennials begin life as a seed but are also able to spread by vegetative reproduction. In vegetative reproduction, plants send out runners known as rhizomes or stolons. These runners are actually horizontal stems. Rhizomes grow under the ground. Stolons grow above the soil surface.

Both rhizomes and stolons give rise to new plants. A plant which spreads by vegetative reproduction can give rise to dozens of new plants. Spreading perennials can take over landscape plantings and large areas of lawn within a few years.

Yellow nutsedge, ground ivy, Canada thistle, hedge bindweed, and quackgrass are some common spreading perennials that spread by rhizomes. Zoysiagrass and bermudagrass are examples of aggressive grasses that spread by stolons. White clover is a broadleaf (non-grass) plant that spreads by stolons.

Life Cycles and Weed Management

Prevention and management of weeds is different for annuals, biennials, and for simple and spreading perennials.

- Annuals - Annual weeds tend to germinate after soil is disturbed. For this reason, they can be a serious problem in new turf or landscape plantings and in established plantings if the soil has been cultivated. Annuals can be controlled by hand pulling, cultivating, burial under mulches, and with herbicides. Annuals are considerably easier to control and to prevent than perennials. Growth of most weeds in new lawns can be controlled with herbicides, mowing, or a combination of the two. In established lawns, growth of annual weeds such as crabgrass can be reduced by reseeding bare patches and improving grass growth so that lawns thicken. In landscapes, annual weeds can be prevented with mulches. Killing annuals by hand pulling is fairly easy. Destruction of the part of the root system just below the soil surface is all that is needed to control annuals, while in perennials the entire root system must be removed.

  The timing of preemergence herbicides is different for winter and summer annuals. Applications of herbicides intended to control summer annuals are made in the spring before weed seedlings emerge. Fall applications of herbicides are sometimes used to control winter annuals. Because of the long emergence period of winter annuals, however, these applications are rarely 100% successful. Preemergence herbicides often disappear from the soil before winter annual weeds have finished emerging.

  Control of annuals may be achieved with postemergence herbicides or contact herbicides. Control of both annuals and perennials can be obtained with systemic (translocated) herbicides such as glyphosate.

- Biennials - Burdock and Queen Anne's lace (wild carrot) are biennials commonly found in New England landscapes. There are few turf weeds which are biennial. Biennials have taproots that must be killed or removed for lasting control. This can be done by hand or with a translocated postemergence herbicide. Growth of these weeds can be prevented with mulches. Biennials spread only by seed and are less likely to be serious problems than many perennials.

- Perennials - Most weeds in established lawns and landscape plantings are perennials. Plantains, dandelion, and ground ivy are examples of perennial turf weeds. Perennial landscape weeds include hedge bindweed, yellow nutsedge, quackgrass (witchgrass) and red sorrel.Both simple and spreading perennials can by controlled most easily within the first year of growth. All portions of the root system must be removed or plants will regrow. For hand control, dig carefully in the ground around the base of the weed, exposing and removing all of the root and rhizomes.

In turf, the growth of many perennial weeds can be controlled with herbicides. Preemergence herbicides kill some weeds as they emerge through soil. Other weeds must be controlled with postemergence materials. Occurrence of turf weeds can be reduced dramatically by maintaining thick, healthy turf. Mowing high (three inches) can help turfgrasses out-compete weeds such as dandelion and plantains.

In landscape beds, both solitary and spreading perennials may also be controlled by directed sprays of glyphosate.

Simple and spreading perennials differ in their importance as weed problems in both landscapes and lawns. Because simple perennials spread only by seed, they can be prevented fairly easily with mulches in landscapes. In lawns, they can be prevented by keeping turf vigorous or by using preemergence herbicides. Spreading perennials are more difficult to control. In landscapes, landscape fabrics will provide partial control of spreading perennials, but well established stands of these weeds tend to come up in gaps around shrubs. Most of the serious weed problems in landscapes are spreading perennials. In lawns, spreading perennials such as yellow nutsedge and quackgrass can form patches in lawns. Because these weeds spread so quickly and are so hard to control, it makes sense to eliminate them before planting begins. When a spreading perennial weed appears in an established planting, control by digging or spot herbicide applications.

References

- Invasive-plant-species: invasivespeciescentre.ca, Retrieved 16 April, 2019

- Non-native-invasive-plants-an-introduction, why-manage-plants: plants.ifas.ufl.edu, Retrieved 19 January, 2019

- Bräutigam, S; Greuter, W (2007). "A new treatment of Pilosella for the Euro-Mediterranean flora". Willdenowia. 37 (1): 123–137. doi:10.3372/wi.37.37106

- Purpleloosestrife, plants-pestmanagement-weedcontrol-noxiouslist: mda.state.mn.us, Retrieved 25 July, 2019

- Villamagna, Amy; Murphy, Brian (27 August 2009). "Ecological and socio-economic impacts of invasive water hyacinth (Eichhornia crassipes): a review". Freshwater Biology. 55 (2): 282–298. doi:10.1111/j.1365-2427.2009.02294.x

- Thijs, Koen W.; Brys, Rein; Verboven, Hans A. F.; Hermy, Martin (30 July 2011). "The influence of an invasive plant species on the pollination success and reproductive output of three riparian plant species". Biological Invasions. 14 (2): 355–365. doi:10.1007/s10530-011-0067-y

Methods of Weed Control

6

Weed control is the method which controls and manages the growth of noxious and invasive weeds. Aquatic weed harvester, stale seed bed, soil steam sterilization, etc. are some of the tools and techniques used for weed control. This chapter discusses about these different tools and related methods of weed control in detail.

WEED CONTROL

Weed control is a botanical component of pest control that stops weeds from reaching a mature stage of growth when they could be harmful to domesticated plants, sometimes livestocks, by using manual techniques including soil cultivation, mulching and herbicides.

Weed control practices in forests are designed to favour the growth of the desired tree species, improve visibility along forest roads, control noxious weeds, and improve wildlife habitats. The goal is to manage timber species, ground vegetation, and wildlife so that each component is maximized yet balanced. Vegetation management is a primary means to achieve a productive forest. Managers need to integrate the best cultural, mechanical, and chemical practices into appropriate and cost effective management systems to minimize losses and detrimental effects due to weeds.

Objectives of Forest Weed Management

A forester might undertake a weed management program with one or more of the following objectives in mind:

Removing unwanted vegetation from planting sites to favor the planted trees. Releasing more desirable species from less desirable overtopping species.Thinning excess plants from a stand. Preventing disease movement through root grafts. Preventing invasion of herbaceous and woody vegetation into recreational areas and wildlife openings. Controlling vegetation along forest roads and around buildings and facilities. Eliminating poisonous plants from recreational areas. Controlling production-limiting weeds in a seed orchard or tree nursery. When establishing a forest, relatively few seeds or seedlings are introduced into an environment in which an almost unlimited number of other plants exist or have the potential to become established. The immediate goal of the forest manager is species survival, which is achieved by reducing the competition from weeds. Site preparation and tree release are the procedures that minimize the density and reduce the vigor of the competing vegetation in the year of and in the years immediately after planting. The type and intensity of management practices depend on the vigor of the desired (planted) species and the indigenous species.

MECHANICAL WEED CONTROL

Mechanical weed control is any physical activity that inhibits unwanted plant growth. Mechanical, or manual, weed control techniques manage weed populations through physical methods that remove, injure, kill, or make the growing conditions unfavorable. Some of these methods cause direct damage to the weeds through complete removal or causing a lethal injury. Other techniques may alter the growing environment by eliminating light, increasing the temperature of the soil, or depriving the plant of carbon dioxide or oxygen. Mechanical control techniques can be either selective or non-selective. A selective method has very little impact on non-target plants where as a non-selective method affects the entire area that is being treated. If mechanical control methods are applied at the optimal time and intensity, some weed species may be controlled or even eradicated.

Mechanical Control Methods

Weed Pulling

Pulling methods uproot and remove the weed from the soil. Weed pulling can be used to control some shrubs, tree saplings, and herbaceous plants. Annuals and tap-rooted weeds tend to be very susceptible to pulling. Many species are able to re-sprout from root segments that are left in the soil. Therefore, the effectiveness of this method is dependent on the removal of as much of the root system as possible. Well established perennial weeds are much less effectively controlled because of the difficulty of removing all of the root system and perennating plant parts. Small herbaceous weeds may be pulled by hand but larger plants may require the use of puller tools like the Weed Wrench or the Root Talon. This technique has a little to no impact on neighboring, non-target plants and has a minimal effect on the growing environment. However, pulling is labor-intensive and time consuming making it a more suitable method to use for small weed infestations.

Mowing

Mowing methods cut or shreds the above ground of the weed and can prevent and reduce seed populations as well as restrict the growth of weeds. Mowing can be a very successful control method for many annual weeds. Mowing is the most effective when it is performed before the weeds are able to set seed because it can reduce the number of flower stalks and prevent the spread of more seed. However, the biology of the weed must be considered before mowing. Some weed species may sprout with increased vigor after being mowed. Also, some species are able to re-sprout from stem or root segments that are left behind after mowing. Brush cutting and weed eating are also mowing techniques that reduce the biomass of the weeds. Repeatedly removing biomass causes reduced vigor in many weed species. This method is usually used in combination with other control methods such as burning or herbicide treatments.

Mulching

Mulch is a layer of material that is spread on the ground. Compared with some other methods of weed control, mulch is relatively simple and inexpensive. Mulching smothers the weeds by excluding light and providing a physical barrier to impede their emergence. Mulching is successful with most annual weeds, however, some perennial weeds are not affected. Mulches may be organic or synthetic. Organic mulches consist of plant by products such as: pine straw, wood chips, green waste, compost, leaves, and grass clippings. Synthetic mulches, also known as ground cover fabric, can be made from materials like polyethylene, polypropylene, or polyester. The effectiveness of mulching is mostly dependent on the material used. Organic and synthetic mulches may be used in combination with each other to increase the amount of weeds controlled.

Tillage

Tillage, also known as cultivation, is the turning over of the soil. This method is more often used in agricultural crops. Tillage can be performed on a small scale with tools such as small, hand pushed rotary tillers or on a large scale with tractor mounted plows. Tillage is able to control weeds because when the soil is overturned, the vegetative parts of the plants are damaged and the root systems are exposed causing desiccation. Generally, the younger the weed is, the more readily it can be controlled with tillage. To control mature perennial weeds, repeated tillage is necessary. By continually destroying new growth and damaging the root system, the weed's food stores are depleted until it can no longer re-sprout. Also, when the soil is overturned, the soil seed bank is disrupted which can cause dormant weed seeds to germinate in the absence of the previous competitors. These new weeds can also be controlled by continued tillage until the soil seed bank is depleted.

Soil Solarization

Soil solarization is a simple method of weed control that is accomplished by covering the soil with a layer of clear or black plastic. The plastic that is covering the ground traps heat energy from the sun and raises the temperature of the soil. Many weed seeds and vegetative propagules are not able to withstand the temperatures and are killed. For this method to be most effective, it should be implemented during the summer months and the soil should be moist. Also, cool season weeds are more susceptible to soil solarization than are warm season weeds. Using black plastic as a cover excludes light which can help to control plants that are growing whereas clear plastic has been shown to produce higher soil temperatures.

Fire

Burning and flaming can be economical and practical methods of weed control if used carefully. For most plants, fire causes the cell walls to rupture when they reach a temperature of 45 °C to 55 °C. Burning is commonly used to control weeds in forests, ditches, and roadsides. Burning can be used to remove accumulated vegetation by destroying the dry, matured plant matter as well as killing the green new growth. Buried weed seeds and plant propagules may also be destroyed during burning, however, dry seeds are much less susceptible to the increased temperature. Flaming is used on a smaller scale and includes the use of a propane torch with a fan tip. Flaming may be used to control weeds along fences and paved areas or places where the soil may be too wet to hoe, dig, or till. Flaming is most effective on young weeds that are less than two inches tall but repeated treatments may control tougher perennial weeds.

Flooding

Flooding is a method of control that requires the area being treated to be saturated at a depth of 15 to 30 cm for a period of 3 to 8 weeks. The saturation of the soil reduces the availability of oxygen to the plant roots thereby killing the weed. This method has been shown to be highly effective in controlling establish perennial weeds and may also suppress annual weeds by reducing the weed seed populations.

Effects of Mechanical Control on the Environment

Mechanical methods of weed control cause physical changes in the immediate environment that may cause positive or negative effects. The suppression of the targeted weeds will open niches in the environment and may also stimulate the growth of other weeds by decreasing their competition and making their environment more favorable. If the niches are not filled by a desirable plant, they will eventually be taken over by another weed. These weed control methods also effect the structure of the soil. The use of mulches can help decrease erosion, decrease water evaporation from the soil, as well as improve the soil structure by increasing the amount of organic matter. Tillage practices can help decrease compaction and aerate the soil. On the other hand, tillage has also been shown to decrease soil moisture, increase soil erosion and runoff, as well as decrease soil microbial populations. Solarization can cause changes in the biological, physical, and chemical properties of the soil. This can cause the soil to be an unfavorable environment for native species which may be beneficial or harmful.

CULTURAL WEED CONTROL

Cultural weed control includes non-chemical crop management practices ranging from variety selection to land preparation to harvest and postharvest processing. Cultural weed control is a part of integrated weed management which involves the integrated use of cultural, manual, and mechanical control methods.

Prepare and level field

Use good clean seed

Plowed fallow field

Maintain plant population

Keeps canals and bunds clean and use water to control weeds

Need for Weed Control

- Prevent yield loss due to weed competition.

- Maintain purity and quality and market price of harvested grain.

- Prevent build-up of weed seeds in soil.

- Prevent weeds that may attract insects or rodents (rats) or act as a host for diseases.

- Prevent clogging of field irrigation channels to facilitate water flow.

- Reduce time and cost of land preparation and weeding operations.

Need for Cultural Control of Weeds

- Cost effective and easy to practice: Acceptable and accessible to small & large farmers.

- Non-chemical and ecologically sound.

- Prevention is better than cure.

Cultural Practices to Control Weeds

- Timing: Weeds need to be controlled from planting until the crop canopy closes.

- Land preparation and leveling: Use land preparation to control growing weeds and to allow weed seeds to germinate. Kill newly emerging weeds by repeat tillage at adequate (~10-day) intervals.

- Reduce weed entry into fields: Prevent the introduction of weeds into fields by: i) use clean good quality seed; ii) keep seedling nurseries free of weeds to make sure weeds are not planted with the rice seedlings; iii) keep irrigation channels and field bunds free of weeds to prevent weed seeds or vegetative parts entering the fields; iv) use clean equipment to prevent field/crop contamination; and v) rotate crops to break weed cycles.

- Fallow management: Kill weeds in fallow fields (e.g., use tillage) to prevent flowering, seed-set and the build-up of weed seeds in the soil (Remember: "one year of seeds, seven years of weeds").

- Crop-weed competition: Select a weed-competitive variety with early seedling vigor, and high tillering to suppress weeds. Transplanted crops tend to have fewer weeds and less yield loss than direct seeded crops. Transplant healthy, vigorous seedlings that can better compete with weeds in early stages. Maintain an adequate plant population that closes its canopy by maximum tillering to shade out weeds. Apply Nitrogen (N) fertilizer just after weeding to minimize rice-weed competition for N.

- Water management: Water is the best control for weeds. Many weeds cannot germinate or grow under flooded conditions (e.g. most grasses and some sedges). Maintain a 2-5 cm. Water level in the field to minimize weed emergence and lower weed pressure. If water is sufficient, fields can be continuously flooded from the time of transplanting to when crop canopy covers the soil completely. Good land leveling is critical to avoid high spots where weeds can become established.

Limitations:

- Needs good knowledge and understanding of various cultural practices that can reduce weed pressure.

- Continuous vigilance and monitoring are essential to control weeds by cultural methods.

- Some practices are labor intensive.

BIOLOGICAL WEED CONTROL

Biological control of weeds is broadly defined as the use of an agent, a complex of agents, or biological processes to bring about weed suppression. All forms of macrobial and microbial organisms are considered as biological control agents. Examples of biological control agents include, but are not limited to: arthropods (insects and mites), plant pathogens (fungi, bacteria, viruses, and nematodes), fish, birds, and other animals. Biologically based weed management is a much broader category of approaches that may include gene modification, genetic processes, and gene products. Human activities intended to remove weeds directly or indirectly, such as hand-weeding and burning, deliberate uses of plant competition, allelopathy, and cultural and soil management practices that alter the biotic balance of soil are considered important adjuncts to biological control in integrated weed management systems.

Underlying Principles and Procedures for Biological Weed Control

Underlying Principles

The underlying principle behind biological approach to weed control is based on some research works that reported that exotic plants become invasive because they have escaped from the insect herbivores and other natural enemies that limit their multiplication and distribution in their native regions; however some other factors may contribute to the tendency for particular plant species to become invasive. Therefore biological control involves using specific natural enemies that can diminish the development and reproduction of their prey organism and put some limitations to them. The predominant approach to classical biological weed control involves the importation, colonization, and establishment of exotic natural enemies (predators, parasites, and pathogens) to diminish and maintain exotic pest populations to densities that are economically insignificant.

General Procedures

Some authors have outlined general procedures to be followed when embarking on classical biological weed control programs as follows: (i) evaluate the ecology, economic impact of the weed and potential conflicts of interest; (ii) survey the organisms that are already attacking the weed in the new habitat in order to distinguish accidentally introduced agents and so eliminate such from future evaluation; (iii) carry out literature search and other forms of survey to identify natural enemies attacking the weed in its native region; (iv) screen the possible biological control agents in the foreign country to determine host range and specificity, and to remove nonspecific agents from further consideration; (v) carry out further tests of promising candidates in quarantine after introduction to ensure host specificity and eliminate predators, parasites, and pathogens that may have been introduced with them; (vi) embark on mass rearing of host-specific agents; (vii) release the host-specific agents; (viii) carry out post-release evaluation to determine establishment and effectiveness of agents; and (ix) redistribute agents to other areas where control is required. Wapshere et al. presented a summary of steps normally followed when introducing a biological control agent in a classical biological control weed program as in table.

Steps	Details
1. Initiation	Data on taxonomy, biology, ecology, economics, native and introduced distributions, known natural enemies, etc., are compiled by initiating scientist or group. An extensive literature review is conducted on the proposed target weed and its relatives, plus known natural enemies. Conflicts of interest identified and resolved if possible
2. Target weed approval	Data in (step 1) submitted to appropriate State and Federal groups for comment; additional data may be required
3. Foreign exploration and domestic surveys	If project approved in (step 2), the center of evolution of the genus of the target weed (if known) and other suitable areas, are searched for natural enemies, particularly where these are eco-climatically similar to the area of introduction. At the same time, the weed should be investigated in the country of introduction for attacking enemies, related plants, etc.
4. Weed ecology and agent host specificity	Ecology of the target weed, its close relatives and its natural enemies is studied in the native area, and the most damaging and apparently selective agents are subjected to several years of host-specificity testing
5. Agent approval	A report on each agent is submitted to appropriate State and Federal bodies to obtain importation and release permits
6. Importation and quarantine clearance	Each agent is imported to the country of introduction where it is reared through at least one generation in quarantine to rid it of its parasites and diseases
7. Rearing and release	After a pure culture of the agent is obtained in (step 6), it is normally mass-reared and released in the field in cages or free at field sites
8. Evaluation and monitoring	Agent is monitored at field sites to determine establishment and degree of stress on target weed, or to determine reasons why the agent did not become established or efficacious
9. Redistribution	To aid spontaneous self-dissemination, agent is distributed to other areas in the target weed's distribution, if needed

Summary of steps normally followed when introducing a biological control agent in a classical/inoculative biological control weed program.

Successful Biological Control of Weeds with Introduced Insects and Pathogens

One thousand one hundred and forty-four individuals (mostly entomologists and plant pathologists) have ever attended the International Symposia on Biological Control of Weeds (ISBCWs); and out of these, 450–550 weed biological control experts have been actively involved in research and development efforts over the last 50 years mainly from USA, Canada, Australia, South Africa and New Zealand. Biological approach to weed control has a long history and a good success rate of 94. Frequently cited examples of successful approach to biological weed control are the prickly pear cacti (*Opuntia; spp.*) in Australia, eradicated by an imported moth (*Cactoblastis cactorum*) and rangeland in California, Oregon, Washington, and British Columbia controlled by St. John's wort *Hypericum perforatum* (millepertuis perforé). Mcfadyen presented a list of 41 weds which have successfully been controlled using introduced insects and pathogens and another three weeds also controlled by introduced fungi applied as mycoherbicides. Many of these successes have been repeated in other countries and continents. Julien presented a list of both successful and failed cases of biological weed control; this included the introduction of 225 organisms against 111 weed species, and 178 insects and 6 mites. new arthropod or pathogen agents were released in 19 projects; and that effective biological control was achieved in several projects with the outstanding successes being the control of rubber vine, *Cryptostegia grandiflora*, and bridal creeper, *Asparagus asparagoides*.

When is Weed Biological Control Successful?

Information collated on weed impacts before the initiation of a biological control program is necessary to provide baseline data and devise performance criteria with which the program can subsequently be evaluated. For avoidance of confusion on when a biological control could be viewed as successful or not, an implementation of a particular biological control will be termed successful when: complete-when no other control method is required or used, at least in areas where the agents is established; substantial-where other methods are needed but the effort required is reduced (e.g. less herbicide or less frequent application); and negligible-where despite damage inflicted by agents, control of the weed is still dependent on other control measures. Complete control does not imply total eradication of the weed; rather it means that control measures are not required anymore specifically against the target weed, and that crop or pasture yield losses will not be attributed mainly to this weed. Substantial control involves situations where control may be complete in some seasons and over part of the weed's range, as well as cases where the control achieved is widespread and economically significant but the weed is still a major problem. It is therefore concluded that successful implementation of biological approach to weed control is the successful control of the weed, and not necessarily the successful establishment of individual agents released against the weed. Successful biological control depends on three factors: the extent to which each individual agent can limit the targeted plant; the ecology of the agent as it affects its ability to populate and spread easily in the new environment; and the ecology of the weed, which determines if the total damage that can be caused by the agent can significantly reduce its population. Because agents always need some surviving predator plants to complete their life cycle, biological control will not usually totally eradicate their target weeds. In essence a successful biological control program reduces the potency and population of the target weed and usually in conjunction with other control methods as part of an overall integrated weed management scheme which is recommended.

Things to Consider when Making the Choice of Agents to be Introduced to Control Weeds

Selection of potential agents in the last decades has been mainly based on the population biology of the weed, impact studies of agents on the plant and the combined effect of herbivory and plant competition. Agent selection is highly dependent on the type of weed, its reproductive system, on the ecological, abiotic and management context in which that weed occurs, and on the acceptable goals and impact thresholds required of a biological control program. Generally, factors to be considered in selecting agents include the following: the agent must target a particular plant species, must have high level of predation and parasitism on the host plant and its entire population, must be prolific, must be able to thrive in all habitats and climates where the weed exists and should be able to spread easily and widely, must be a strong colonizer, the overall cost of introducing the agent must be cheaper compared to other control methods, the technology that will be involved in introducing and managing the agent must be as simple as possible, must as much as possible maintain natural biodiversity, sufficient number of individuals must be released, plant phenology (effect of periodic plant life cycle events) must be favorable. To be considered a good candidate for biological control, a weed should be non-native, present in numbers and densities greater than in its native range and numerous enough to cause environmental or economic damage, the weed should also be present over a broad geographic range, have few or no redeeming or beneficial qualities, have taxonomic characteristics sufficiently distinct from those of economically important and native plant. Furthermore, the weed should occur in relatively undisturbed areas to allow for the establishment of biological control agents, cultivation, mowing and other disturbances can have a destructive effect on many arthropod biocontrol agents. Inundative biocontrol agents such as bacteria and fungi are less sensitive to these types of disturbances so may be used in cropland.

Steps to Identifying and Introducing Biological Control Agents

The study of insect attributes and fitness traits, the influence of plant resources on insect performance, and the construction of comparative life-tables, are the first steps towards an improvement of the success rate of biological weed control. Generally, steps to identifying and introducing biological control agents include: (i) identify target weeds; (ii) identify control agents and determine the level of specialization; (iii) apply controlled release of the agents; (iv) apply full release and determine optimal release sites; (v) for the case of classical methods, monitor release sites; (vi) apply redistribution for the case of classical methods (vii) and maintain control agent populations.

CHEMICAL WEED CONTROL

In chemical weed control, chemicals called herbicides are used to kill certain plants or inhibit their growth. Chemical weed control is an option in integrated weed management that refers to the integrated use of cultural, manual, mechanical and chemical control methods.

Chemical weed control has been used for a very long time: sea salt, industrial by-products, and oils were first employed. Selective control of broad-leaved weeds in fields of cereal crops was discovered in France in the late 1800s, and this practice soon spread throughout Europe. Sulfates and nitrates of copper and iron were used; sulfuric acid proved even more effective. Application was by spraying. Soon sodium arsenite became popular both as a spray and as a soil sterilant. On thousands of miles of railroad right-of-way, and in sugarcane and rubber plantations in the tropics, the hazardous material was used in tremendous quantities, often resulting in the poisoning of animals and occasionally humans. Diesel oil, as a general herbicide, and sodium dinitrocresylate (Sinox), as a selective plant killer, were introduced during the first three decades of the 20th century.

Spraying herbicides on cereal crops.

Sinox, the first major organic chemical herbicide, was developed in France in 1896. In the late 1940s new herbicides were developed out of the research during World War II, and the era of the miracle weed killers began. Within 20 years over 100 new chemicals were synthesized, developed, and put into use. Chemical weed control superseded both plant-disease and insect-pest control in economic impact. The year 1945 marked the beginning of a new era in chemical weed control. Introduced then were 2,4-D (2,4-dichlorophenoxyacetic acid), 2,4,5-T (2,4,5-trichlorophenoxy-acetic acid), and IPC (isopropyl-N-phenylcarbamate), the first two selective as foliar sprays against broad-leaved weeds, the third selective against grass species when applied through the soil. The new herbicides were revolutionary in that their high toxicity allowed for effective weed control at dosage rates as low as one to two pounds per acre. This contrasted with carbon bisulfide, borax, and arsenic trioxide, which were required at rates of up to one ton per acre, and with sodium chlorate, required at rates of around 100 pounds per acre. However, some of those early herbicides, including 2,4,5-T, were later deemed unsafe for humans and the environment and were discontinued in many countries. Effective herbicides have continued to be developed, and some, such as glyphosate, are widely used around the world.

Herbicides may be grouped into two categories: selective and nonselective. Each category may be subdivided into foliage-applied and soil-applied materials and, in cases where field crops are treated, the application may be made before sowing the crop (pre-planting), after sowing but before emergence of seedlings (pre-emergence), or after seedlings have emerged (post-emergence).

A great advantage of chemical over mechanical weed control is the ease of application. This is particularly true in cereal croplands, pastures, rangelands, forests, and other situations where an airplane can be used. Many millions of acres are treated from the air each year, and modern equipment for treating row-crop land with herbicides has made weed control increasingly convenient. Sprayers, soil incorporation equipment, and spreaders for pelleted herbicides have all added to the convenience of, and removed uncertainty from, herbicide application. Machinery is available that simultaneously builds up beds, plants the seed, sprays with insecticide, and incorporates fertilizer and pre-emergence herbicide all in one operation.

Certain agricultural plants, known as herbicide-resistant crops (HRCs), have been genetically engineered for resistance to specific chemical herbicides, notably glyphosate. These genetically modified organisms (GMOs) have been available since the mid-1980s and enable effective chemical control of weeds, since only the HRC plants can survive in fields treated with the corresponding herbicide. However, because these crops encourage increased application of chemicals to the soil rather than decreased application, they remain controversial with regard to their environmental impact and general safety.

Genetically modified (GM) barley grown by researchers.
The GM barley was investigated for its effects on soil quality.

HERBICIDES

Herbicides also commonly known as weedkillers, are substances used to control unwanted plants. Selective herbicides control specific weed species, while leaving the desired crop relatively unharmed, while non-selective herbicides (sometimes called total weedkillers in commercial products) can be used to clear waste ground, industrial and construction sites, railways and railway embankments as they kill all plant material with which they come into contact. Apart from selective/non-selective, other important distinctions include *persistence* (also known as *residual action*: how long the product stays in place and remains active), *means of uptake* (whether it is absorbed by above-ground foliage only, through the roots, or by other means), and *mechanism of action* (how it works). Historically, products such as common salt and other metal salts were used as herbicides, however these have gradually fallen out of favor and in some countries a number of these are banned due to their persistence in soil, and toxicity and groundwater contamination concerns. Herbicides have also been used in warfare and conflict.

Weeds controlled with herbicide.

Modern herbicides are often synthetic mimics of natural plant hormones which interfere with growth of the target plants. The term organic herbicide has come to mean herbicides intended for organic farming. Some plants also produce their own natural herbicides, such as the genus *Juglans* (walnuts), or the tree of heaven; such action of natural herbicides, and other related chemical interactions, is called allelopathy. Due to herbicide resistance - a major concern in agriculture - a number of products combine herbicides with different means of action. Integrated pest management may use herbicides alongside other pest control methods.

In the US in 2007, about 83% of all herbicide usage, determined by weight applied, was in agriculture. In 2007, world pesticide expenditures totaled about $39.4 billion; herbicides were about 40% of those sales and constituted the biggest portion, followed by insecticides, fungicides, and other types. Herbicide is also used in forestry, where certain formulations have been found to suppress hardwood varieties in favour of conifers after a clearcut, as well as pasture systems, and management of areas set aside as wildlife habitat.

Herbicides are classified/grouped in various ways e.g. according to the activity, timing of application, method of application, mechanism of action, chemical family. This gives rise to a considerable level of terminology related to herbicides and their use.

Intended Outcome

- Control is the destruction of unwanted weeds, or the damage of them to the point where they are no longer competitive with the crop.

- Suppression is incomplete control still providing some economic benefit, such as reduced competition with the crop.

- Crop safety, for selective herbicides, is the relative absence of damage or stress to the crop. Most selective herbicides cause some visible stress to crop plants.

- Defoliant, similar to herbicides, but designed to remove foliage (leaves) rather than kill the plant.

Selectivity

- Selective herbicides control or suppress certain plants without affecting the growth of other plants species. Selectivity may be due to translocation, differential absorption, physical (morphological) or physiological differences between plant species. 2,4-D, mecoprop, dicamba control many broadleaf weeds but remain ineffective against turfgrasses.

- Non-selective herbicides are not specific in acting against certain plant species and control all plant material with which they come into contact. They are used to clear industrial sites, waste ground, railways and railway embankments. Paraquat, glufosinate, glyphosate are non-selective herbicides.

Timing of Application

- Pre-plant: Pre-plant herbicides are non-selective herbicides applied to soil before planting. Some pre-plant herbicides may be mechanically incorporated into the soil. The objective for incorporation is to prevent dissipation through photo-decomposition and volatility. The herbicides kill weeds as they grow through the herbicide treated zone. Volatile herbicides have to be incorporated into the soil before planting the pasture. Agricultural crops grown in soil treated with a pre-plant herbicide include tomatoes, corn, soybeans and strawberries. Soil fumigants like metam-sodium and dazomet are in use as pre-plant herbicides.

- Pre-emergence: Pre-emergence herbicides are applied before the weed seedlings emerge through the soil surface. Herbicides do not prevent weeds from germinating but they kill weeds as they grow through the herbicide treated zone by affecting the cell division in the emerging seedling. Dithopyr and pendimethalin are pre-emergence herbicides. Weeds that have already emerged before application or activation are not affected by pre-herbicides as their primary growing point escapes the treatment.

- Post-emergence: These herbicides are applied after weed seedlings have emerged through the soil surface. They can be foliar or root absorbed, selective or non-selective, contact or systemic. Application of these herbicides is avoided during rain because the problem of being washed off to the soil makes it ineffective. 2,4-D is a selective, systemic, foliar absorbed post-emergence herbicide.

Method of Application

- Soil applied: Herbicides applied to the soil are usually taken up by the root or shoot of the emerging seedlings and are used as pre-plant or pre-emergence treatment. Several factors influence the effectiveness of soil-applied herbicides. Weeds absorb herbicides by both passive and active mechanism. Herbicide adsorption to soil colloids or organic matter often reduces its amount available for weed absorption. Positioning of herbicide in correct layer of soil is very important, which can be achieved mechanically and by rainfall. Herbicides on the soil surface are subjected to several processes that reduce their availability. Volatility and photolysis are two common processes that reduce the availability of herbicides. Many soil applied herbicides are absorbed through plant shoots while they are still underground leading to their death or injury. EPTC and trifluralin are soil applied herbicides.

- Foliar applied: These are applied to portion of the plant above the ground and are absorbed by exposed tissues. These are generally post-emergence herbicides and can either be translocated (systemic) throughout the plant or remain at specific site (contact). External barriers of plants like cuticle, waxes, cell wall etc. affect herbicide absorption and action. Glyphosate, 2,4-D and dicamba are foliar applied herbicide.

Persistence

- Residual activity: An herbicide is described as having low residual activity if it is neutralized within a short time of application (within a few weeks or months) - typically this is due to rainfall, or by reactions in the soil. An herbicide described as having high residual activity will remain potent for a long term in the soil. For some compounds, the residual activity can leave the ground almost permanently barren.

Mechanism of Action

Herbicides are often classified according to their site of action, because as a general rule, herbicides within the same site of action class will produce similar symptoms on susceptible plants. Classification based on site of action of herbicide is comparatively better as herbicide resistance management can be handled more properly and effectively. Classification by mechanism of action (MOA) indicates the first enzyme, protein, or biochemical step affected in the plant following application.

List of Mechanisms found in Modern Herbicides:

- ACCase inhibitors: Acetyl coenzyme A carboxylase (ACCase) is part of the first step of lipid synthesis. Thus, ACCase inhibitors affect cell membrane production in the meristems of the grass plant. The ACCases of grasses are sensitive to these herbicides, whereas the AC-Cases of dicot plants are not.

- ALS inhibitors: the acetolactate synthase (ALS) enzyme (also known as acetohydroxyacid synthase, or AHAS) is the first step in the synthesis of the branched-chain amino acids (valine, leucine, and isoleucine). These herbicides slowly starve affected plants of these amino acids, which eventually leads to inhibition of DNA synthesis. They affect grasses and dicots alike. The ALS inhibitor family includes various sulfonylureas (SUs) (such as Flazasulfuron and Metsulfuron-methyl), imidazolinones (IMIs), triazolopyrimidines (TPs), pyrimidinyl oxybenzoates (POBs), and sulfonylamino carbonyl triazolinones (SCTs). The ALS biological pathway exists only in plants and not animals, thus making the ALS-inhibitors among the safest herbicides.

- EPSPS inhibitors: Enolpyruvylshikimate 3-phosphate synthase enzyme (EPSPS) is used in the synthesis of the amino acids tryptophan, phenylalanine and tyrosine. They affect grasses and dicots alike. Glyphosate (Roundup) is a systemic EPSPS inhibitor inactivated by soil contact.

- The discovery of synthetic auxins inaugurated the era of organic herbicides. They were discovered in the 1940s after long study of the plant growth regulator auxin. Synthetic auxins mimic in some way this plant hormone. They have several points of action on the cell membrane, and are effective in the control of dicot plants. 2,4-D and 2,4,5-T are synthetic auxin herbicides.

- Photosystem II inhibitors reduce electron flow from water to $NADP^+$ at the photochemical step in photosynthesis. They bind to the Qb site on the D1 protein, and prevent quinone from binding to this site. Therefore, this group of compounds causes electrons to accumulate on chlorophyll molecules. As a consequence, oxidation reactions in excess of those normally tolerated by the cell occur, and the plant dies. The triazine herbicides (including atrazine) and urea derivatives (diuron) are photosystem II inhibitors.

- Photosystem I inhibitors steal electrons from the normal pathway through FeS to Fdx to $NADP^+$ leading to direct discharge of electrons on oxygen. As a result, reactive oxygen species are produced and oxidation reactions in excess of those normally tolerated by the cell occur, leading to plant death. Bipyridinium herbicides (such as diquat and paraquat) inhibit the FeS to Fdx step of that chain, while diphenyl ether herbicides (such as nitrofen, nitrofluorfen, and acifluorfen) inhibit the Fdx to $NADP^+$ step.

- HPPD inhibitors inhibit 4-Hydroxyphenylpyruvate dioxygenase, which are involved in tyrosine breakdown. Tyrosine breakdown products are used by plants to make carotenoids, which protect chlorophyll in plants from being destroyed by sunlight. If this happens, the plants turn white due to complete loss of chlorophyll, and the plants die. Mesotrione and sulcotrione are herbicides in this class; a drug, nitisinone, was discovered in the course of developing this class of herbicides.

Herbicide Group

One of the most important methods for preventing, delaying, or managing resistance is to reduce the reliance on a single herbicide mode of action. To do this, farmers must know the mode of action for the herbicides they intend to use, but the relatively complex nature of plant biochemistry makes this difficult to determine. Attempts were made to simplify the understanding of herbicide mode of action by developing a classification system that grouped herbicides by mode of action. Eventually the Herbicide Resistance Action Committee (HRAC) and the Weed Science Society of America (WSSA) developed a classification system. The WSSA and HRAC systems differ in the group designation. Groups in the WSSA and the HRAC systems are designated by numbers and letters, respectively. The goal for adding the "Group" classification and mode of action to the herbicide product label is to provide a simple and practical approach to deliver the information to users. This information will make it easier to develop educational material that is consistent and effective. It should increase user's awareness of herbicide mode of action and provide more accurate recommendations for resistance management. Another goal is to make it easier for users to keep records on which herbicide mode of actions are being used on a particular field from year to year.

Chemical Family

Detailed investigations on chemical structure of the active ingredients of the registered herbicides showed that some moieties (moiety is a part of a molecule that may include either whole functional groups or parts of functional groups as substructures; a functional group has similar chemical properties whenever it occurs in different compounds) have the same mechanisms of action. According to Forouzesh *et al.*, these moieties have been assigned to the names of chemical families and active ingredients are then classified within the chemical families accordingly. Knowing about herbicide chemical family grouping could serve as a short-term strategy for managing resistance to site of action.

Use and Application

Herbicides being sprayed from the spray arms of a tractor in North Dakota.

Most herbicides are applied as water-based sprays using ground equipment. Ground equipment varies in design, but large areas can be sprayed using self-propelled sprayers equipped with long booms, of 60 to 120 feet (18 to 37 m) with spray nozzles spaced every 20–30 inches (510–760 mm) apart. Towed, handheld, and even horse-drawn sprayers are also used. On large areas, herbicides may also at times be applied aerially using helicopters or airplanes, or through irrigation systems (known as chemigation).

A further method of herbicide application developed around 2010, involves ridding the soil of its active weed seed bank rather than just killing the weed. This can successfully treat annual plants but not perennials. Researchers at the Agricultural Research Service found that the application of herbicides to fields late in the weeds' growing season greatly reduces their seed production, and therefore fewer weeds will return the following season. Because most weeds are annuals, their seeds will only survive in soil for a year or two, so this method will be able to destroy such weeds after a few years of herbicide application.

Weed-wiping may also be used, where a wick wetted with herbicide is suspended from a boom and dragged or rolled across the tops of the taller weed plants. This allows treatment of taller grassland weeds by direct contact without affecting related but desirable shorter plants in the grassland sward beneath. The method has the benefit of avoiding spray drift. In Wales, a scheme offering free weed-wiper hire was launched in 2015 in an effort to reduce the levels of MCPA in water courses.

Health and Environmental Effects

Herbicides have widely variable toxicity in addition to acute toxicity arising from ingestion of a significant quantity rapidly, and chronic toxicity arising from environmental and occupational exposure over long periods. Much public suspicion of herbicides revolves around a confusion between valid statements of *acute* toxicity as opposed to equally valid statements of lack of *chronic* toxicity at the recommended levels of usage. For instance, while glyphosate formulations with tallowamine *adjuvants* are acutely toxic, their use was found to be uncorrelated with any health issues like cancer in a massive US Department of Health study on 90,000 members of farmer families for over a period of 23 years. That is, the study shows lack of chronic toxicity, but cannot question the herbicide's acute toxicity.

Some herbicides cause a range of health effects ranging from skin rashes to death. The pathway of attack can arise from intentional or unintentional direct consumption, improper application resulting in the herbicide coming into direct contact with people or wildlife, inhalation of aerial sprays, or food consumption prior to the labelled preharvest interval. Under some conditions, certain herbicides can be transported via leaching or surface runoff to contaminate groundwater or distant surface water sources. Generally, the conditions that promote herbicide transport include intense storm events (particularly shortly after application) and soils with limited capacity to adsorb or retain the herbicides. Herbicide properties that increase likelihood of transport include persistence (resistance to degradation) and high water solubility.

Phenoxy herbicides are often contaminated with dioxins such as TCDD; research has suggested such contamination results in a small rise in cancer risk after occupational exposure to these herbicides. Triazine exposure has been implicated in a likely relationship to increased risk of breast cancer, although a causal relationship remains unclear.

Herbicide manufacturers have at times made false or misleading claims about the safety of their products. Chemical manufacturer Monsanto Company agreed to change its advertising after pressure from New York attorney general Dennis Vacco; Vacco complained about misleading claims that its spray-on glyphosate-based herbicides, including Roundup, were safer than table salt and "practically non-toxic" to mammals, birds, and fish (though proof that this was ever said is hard to find). Roundup is toxic and has resulted in death after being ingested in quantities ranging from 85 to 200 ml, although it has also been ingested in quantities as large as 500 ml with only mild or moderate symptoms. The manufacturer of Tordon 101 (Dow AgroSciences, owned by the Dow Chemical Company) has claimed Tordon 101 has no effects on animals and insects, in spite of evidence of strong carcinogenic activity of the active ingredient Picloram in studies on rats.

The risk of Parkinson's disease has been shown to increase with occupational exposure to herbicides and pesticides. The herbicide paraquat is suspected to be one such factor.

All commercially sold, organic and nonorganic herbicides must be extensively tested prior to approval for sale and labeling by the Environmental Protection Agency. However, because of the large number of herbicides in use, concern regarding health effects is significant. In addition to health effects caused by herbicides themselves, commercial herbicide mixtures often contain other chemicals, including inactive ingredients, which have negative impacts on human health.

Ecological Effects

Commercial herbicide use generally has negative impacts on bird populations, although the impacts are highly variable and often require field studies to predict accurately. Laboratory studies have at times overestimated negative impacts on birds due to toxicity, predicting serious problems that were not observed in the field. Most observed effects are due not to toxicity, but to habitat changes and the decreases in abundance of species on which birds rely for food or shelter. Herbicide use in silviculture, used to favor certain types of growth following clearcutting, can cause significant drops in bird populations. Even when herbicides which have low toxicity to birds are used, they decrease the abundance of many types of vegetation on which the birds rely. Herbicide use in agriculture in Britain has been linked to a decline in seed-eating bird species which rely on the weeds killed by the herbicides. Heavy use of herbicides in neotropical agricultural areas has been one of many factors implicated in limiting the usefulness of such agricultural land for wintering migratory birds.

Frog populations may be affected negatively by the use of herbicides as well. While some studies have shown that atrazine may be a teratogen, causing demasculinization in male frogs, the U.S. Environmental Protection Agency (EPA) and its independent Scientific Advisory Panel (SAP) examined all available studies on this topic and concluded that "atrazine does not adversely affect amphibian gonadal development based on a review of laboratory and field studies."

Scientific Uncertainty of Full Extent of Herbicide Effects

The health and environmental effects of many herbicides is unknown, and even the scientific community often disagrees on the risk. For example, a 1995 panel of 13 scientists reviewing studies on the carcinogenicity of 2,4-D had divided opinions on the likelihood 2,4-D causes cancer in humans. As of 1992, studies on phenoxy herbicides were too few to accurately assess the risk of many types of cancer from these herbicides, even though evidence was stronger that exposure to these herbicides is associated with increased risk of soft tissue sarcoma and non-Hodgkin lymphoma. Furthermore, there is some suggestion that herbicides can play a role in sex reversal of certain organisms that experience temperature-dependent sex determination, which could theoretically alter sex ratios.

Resistance

Weed resistance to herbicides has become a major concern in crop production worldwide. Resistance to herbicides is often attributed to lack of rotational programmes of herbicides and to continuous applications of herbicides with the same sites of action. Thus, a true understanding of the sites of action of herbicides is essential for strategic planning of herbicide-based weed control.

Plants have developed resistance to atrazine and to ALS-inhibitors, and more recently, to glyphosate herbicides. Marestail is one weed that has developed glyphosate resistance. Glyphosate-resistant weeds are present in the vast majority of soybean, cotton and corn farms in some U.S. states. Weeds that can resist multiple other herbicides are spreading. Few new herbicides are near commercialization, and none with a molecular mode of action for which there is no resistance. Because most herbicides could not kill all weeds, farmers rotated crops and herbicides to stop resistant weeds. During its initial years, glyphosate was not subject to resistance and allowed farmers to reduce the use of rotation.

A family of weeds that includes waterhemp (Amaranthus rudis) is the largest concern. A 2008-9 survey of 144 populations of waterhemp in 41 Missouri counties revealed glyphosate resistance in 69%. Weeds from some 500 sites throughout Iowa in 2011 and 2012 revealed glyphosate resistance in approximately 64% of waterhemp samples. The use of other killers to target "residual" weeds has become common, and may be sufficient to have stopped the spread of resistance From 2005 through 2010 researchers discovered 13 different weed species that had developed resistance to glyphosate. But since then only two more have been discovered. Weeds resistant to multiple herbicides with completely different biological action modes are on the rise. In Missouri, 43% of samples were resistant to two different herbicides; 6% resisted three; and 0.5% resisted four. In Iowa 89% of waterhemp samples resist two or more herbicides, 25% resist three, and 10% resist five.

For southern cotton, herbicide costs has climbed from between $50 and $75 per hectare a few years ago to about $370 per hectare in 2013. Resistance is contributing to a massive shift away from growing cotton; over the past few years, the area planted with cotton has declined by 70% in Arkansas and by 60% in Tennessee. For soybeans in Illinois, costs have risen from about $25 to $160 per hectare.

Dow, Bayer CropScience, Syngenta and Monsanto are all developing seed varieties resistant to herbicides other than glyphosate, which will make it easier for farmers to use alternative weed killers. Even though weeds have already evolved some resistance to those herbicides, Powles says the new seed-and-herbicide combos should work well if used with proper rotation.

Biochemistry of Resistance

Resistance to herbicides can be based on one of the following biochemical mechanisms:

- Target-site resistance: This is due to a reduced (or even lost) ability of the herbicide to bind to its target protein. The effect usually relates to an enzyme with a crucial function in a metabolic pathway, or to a component of an electron-transport system. Target-site resistance may also be caused by an overexpression of the target enzyme (via gene amplification or changes in a gene promoter).

- Non-target-site resistance: This is caused by mechanisms that reduce the amount of herbicidal active compound reaching the target site. One important mechanism is an enhanced metabolic detoxification of the herbicide in the weed, which leads to insufficient amounts of the active substance reaching the target site. A reduced uptake and translocation, or sequestration of the herbicide, may also result in an insufficient herbicide transport to the target site.

- Cross-resistance: In this case, a single resistance mechanism causes resistance to several herbicides. The term target-site cross-resistance is used when the herbicides bind to the same target site, whereas non-target-site cross-resistance is due to a single non-target-site mechanism (e.g., enhanced metabolic detoxification) that entails resistance across herbicides with different sites of action.

- Multiple resistance: In this situation, two or more resistance mechanisms are present within individual plants, or within a plant population.

Resistance Management

Worldwide experience has been that farmers tend to do little to prevent herbicide resistance developing, and only take action when it is a problem on their own farm or neighbor's. Careful observation is important so that any reduction in herbicide efficacy can be detected. This may indicate evolving resistance. It is vital that resistance is detected at an early stage as if it becomes an acute, whole-farm problem, options are more limited and greater expense is almost inevitable. Table lists factors which enable the risk of resistance to be assessed. An essential pre-requisite for confirmation of resistance is a good diagnostic test. Ideally this should be rapid, accurate, cheap and accessible. Many diagnostic tests have been developed, including glasshouse pot assays, petri dish assays and chlorophyll fluorescence. A key component of such tests is that the response of the suspect population to a herbicide can be compared with that of known susceptible and resistant standards under controlled conditions. Most cases of herbicide resistance are a consequence of the repeated use of herbicides, often in association with crop monoculture and reduced cultivation practices. It is necessary, therefore, to modify these practices in order to prevent or delay the onset of resistance or to control existing resistant populations. A key objective should be the reduction in selection pressure. An integrated weed management (IWM) approach is required, in which as many tactics as possible are used to combat weeds. In this way, less reliance is placed on herbicides and so selection pressure should be reduced.

Optimising herbicide input to the economic threshold level should avoid the unnecessary use of herbicides and reduce selection pressure. Herbicides should be used to their greatest potential by ensuring that the timing, dose, application method, soil and climatic conditions are optimal for good activity. In the UK, partially resistant grass weeds such as *Alopecurus myosuroides* (blackgrass) and *Avena* spp. (wild oat) can often be controlled adequately when herbicides are applied at the 2-3 leaf stage, whereas later applications at the 2-3 tiller stage can fail badly. Patch spraying, or applying herbicide to only the badly infested areas of fields, is another means of reducing total herbicide use.

Table: Agronomic factors influencing the risk of herbicide resistance development.

Factor	Low risk	High risk
Cropping system	Good rotation	Crop monoculture
Cultivation system	Annual ploughing	Continuous minimum tillage
Weed control	Cultural only	Herbicide only
Herbicide use	Many modes of action	Single modes of action
Control in previous years	Excellent	Poor
Weed infestation	Low	High
Resistance in vicinity	Unknown	Common

Approaches to Treating Resistant Weeds

Alternative Herbicides

When resistance is first suspected or confirmed, the efficacy of alternatives is likely to be the first consideration. The use of alternative herbicides which remain effective on resistant populations can be a successful strategy, at least in the short term. The effectiveness of alternative herbicides will be highly dependent on the extent of cross-resistance. If there is resistance to a single group of herbicides, then the use of herbicides from other groups may provide a simple and effective solution, at least in the short term. For example, many triazine-resistant weeds have been readily controlled by the use of alternative herbicides such as dicamba or glyphosate. If resistance extends to more than one herbicide group, then choices are more limited. It should not be assumed that resistance will automatically extend to all herbicides with the same mode of action, although it is wise to assume this until proved otherwise. In many weeds the degree of cross-resistance between the five groups of ALS inhibitors varies considerably. Much will depend on the resistance mechanisms present, and it should not be assumed that these will necessarily be the same in different populations of the same species. These differences are due, at least in part, to the existence of different mutations conferring target site resistance. Consequently, selection for different mutations may result in different patterns of cross-resistance. Enhanced metabolism can affect even closely related herbicides to differing degrees. For example, populations of *Alopecurus myosuroides* (blackgrass) with an enhanced metabolism mechanism show resistance to pendimethalin but not to trifluralin, despite both being dinitroanilines. This is due to differences in the vulnerability of these two herbicides to oxidative metabolism. Consequently, care is needed when trying to predict the efficacy of alternative herbicides.

Mixtures and Sequences

The use of two or more herbicides which have differing modes of action can reduce the selection for resistant genotypes. Ideally, each component in a mixture should:

- Be active at different target sites.
- Have a high level of efficacy.
- Be detoxified by different biochemical pathways.
- Have similar persistence in the soil (if it is a residual herbicide).
- Exert negative cross-resistance.
- Synergise the activity of the other component.

No mixture is likely to have all these attributes, but the first two listed are the most important. There is a risk that mixtures will select for resistance to both components in the longer term. One practical advantage of sequences of two herbicides compared with mixtures is that a better appraisal of the efficacy of each herbicide component is possible, provided that sufficient time elapses between each application. A disadvantage with sequences is that two separate applications have to be made and it is possible that the later application will be less effective on weeds surviving the first application. If these are resistant, then the second herbicide in the sequence may increase selection for resistant individuals by killing the susceptible plants which were damaged but not killed by the first application, but allowing the larger, less affected, resistant plants to survive. This has been cited as one reason why ALS-resistant *Stellaria media* has evolved in Scotland recently, despite the regular use of a sequence incorporating mecoprop, a herbicide with a different mode of action.

Herbicide Rotations

Rotation of herbicides from different chemical groups in successive years should reduce selection for resistance. This is a key element in most resistance prevention programmes. The value of this approach depends on the extent of cross-resistance, and whether multiple resistance occurs owing to the presence of several different resistance mechanisms. A practical problem can be the lack of awareness by farmers of the different groups of herbicides that exist. In Australia a scheme has been introduced in which identifying letters are included on the product label as a means of enabling farmers to distinguish products with different modes of action.

List of Common Herbicides

Synthetic Herbicides

- 2,4-D is a broadleaf herbicide in the phenoxy group used in turf and no-till field crop production. Now, it is mainly used in a blend with other herbicides to allow lower rates of herbicides to be used; it is the most widely used herbicide in the world, and third most commonly used in the United States. It is an example of synthetic auxin (plant hormone).

- Aminopyralid is a broadleaf herbicide in the pyridine group, used to control weeds on grassland, such as docks, thistles and nettles. It is notorious for its ability to persist in compost.

- Atrazine, a triazine herbicide, is used in corn and sorghum for control of broadleaf weeds and grasses. Still used because of its low cost and because it works well on a broad spectrum of weeds common in the US corn belt, atrazine is commonly used with other herbicides to reduce the overall rate of atrazine and to lower the potential for groundwater contamination; it is a photosystem II inhibitor.

- Clopyralid is a broadleaf herbicide in the pyridine group, used mainly in turf, rangeland, and for control of noxious thistles. Notorious for its ability to persist in compost, it is another example of synthetic auxin.

- Dicamba, a postemergent broadleaf herbicide with some soil activity, is used on turf and field corn. It is another example of a synthetic auxin.

- Glufosinate ammonium, a broad-spectrum contact herbicide, is used to control weeds after the crop emerges or for total vegetation control on land not used for cultivation.

- Fluazifop (Fuselade Forte), a post emergence, foliar absorbed, translocated grass-selective herbicide with little residual action. It is used on a very wide range of broad leaved crops for control of annual and perennial grasses.

- Fluroxypyr, a systemic, selective herbicide, is used for the control of broad-leaved weeds in small grain cereals, maize, pastures, rangeland and turf. It is a synthetic auxin. In cereal growing, fluroxypyr's key importance is control of cleavers, *Galium aparine*. Other key broadleaf weeds are also controlled.

- Glyphosate, a systemic nonselective herbicide, is used in no-till burndown and for weed control in crops genetically modified to resist its effects. It is an example of an EPSPs inhibitor.

- Imazapyr a nonselective herbicide, is used for the control of a broad range of weeds, including terrestrial annual and perennial grasses and broadleaf herbs, woody species, and riparian and emergent aquatic species.

- Imazapic, a selective herbicide for both the pre- and postemergent control of some annual and perennial grasses and some broadleaf weeds, kills plants by inhibiting the production of branched chain amino acids (valine, leucine, and isoleucine), which are necessary for protein synthesis and cell growth.

- Imazamox, an imidazolinone manufactured by BASF for post-emergence application that is an acetolactate synthase (ALS) inhibitor. Sold under trade names Raptor, Beyond, and Clearcast.

- Linuron is a nonselective herbicide used in the control of grasses and broadleaf weeds. It works by inhibiting photosynthesis.

- MCPA (2-methyl-4-chlorophenoxyacetic acid) is a phenoxy herbicide selective for broadleaf plants and widely used in cereals and pasture.

- Metolachlor is a pre-emergent herbicide widely used for control of annual grasses in corn and sorghum; it has displaced some of the atrazine in these uses.

- Paraquat is a nonselective contact herbicide used for no-till burndown and in aerial destruction of marijuana and coca plantings. It is more acutely toxic to people than any other herbicide in widespread commercial use.

- Pendimethalin, a pre-emergent herbicide, is widely used to control annual grasses and some broad-leaf weeds in a wide range of crops, including corn, soybeans, wheat, cotton, many tree and vine crops, and many turfgrass species.

- Picloram, a pyridine herbicide, mainly is used to control unwanted trees in pastures and edges of fields. It is another synthetic auxin.

- Sodium chlorate *(disused/banned in some countries)*, a nonselective herbicide, is considered phytotoxic to all green plant parts. It can also kill through root absorption.

- Triclopyr, a systemic, foliar herbicide in the pyridine group, is used to control broadleaf weeds while leaving grasses and conifers unaffected.

- Several sulfonylureas, including Flazasulfuron and Metsulfuron-methyl, which act as ALS inhibitors and in some cases are taken up from the soil via the roots.

Organic Herbicides

Recently, the term "organic" has come to imply products used in organic farming. Under this definition, an organic herbicide is one that can be used in a farming enterprise that has been classified as organic. Depending on the application, they may be less effective than synthetic herbicides and are generally used along with cultural and mechanical weed control practices.

Homemade organic herbicides include:

- Corn gluten meal (CGM) is a natural pre-emergence weed control used in turfgrass, which reduces germination of many broadleaf and grass weeds.

- Vinegar is effective for 5–20% solutions of acetic acid, with higher concentrations most effective, but it mainly destroys surface growth, so respraying to treat regrowth is needed. Resistant plants generally succumb when weakened by respraying.

- Steam has been applied commercially, but is now considered uneconomical and inadequate. It controls surface growth but not underground growth and so respraying to treat regrowth of perennials is needed.

- Flame is considered more effective than steam, but suffers from the same difficulties.

- D-limonene (citrus oil) is a natural degreasing agent that strips the waxy skin or cuticle from weeds, causing dehydration and ultimately death.

- Saltwater or salt applied in appropriate strengths to the rootzone will kill most plants.

Pre-emergent Herbicide

Pre-emergent herbicides prevent the germination of seeds by inhibiting a key enzyme. In some areas of the world, they are used to prevent crabgrass from appearing in lawns. Pre-emergent herbicides are applied to lawns in the spring and fall, to prevent the germination of weed seeds. They will not affect any established plant. In the spring, they should be applied when air temperatures reach 65–70°F for four consecutive days. In the fall, they should be applied when nighttime lows reach 55–60°F for four consecutive nights.

A pre-emergence spray of herbicide being added onto a field of oilseed rape.

"Weed and feed" products which contain both Pre-emergent herbicide and fertilizer in a single product should not be used on southern lawns or warm-season grasses. If applied when Pre-emergent herbicide is needed, the fertilizer may burn or stress the lawn. If applied after the lawn "green-up", weed seeds will have already germinated and the herbicide will be ineffective.

Crabgrass

To prevent growth of crabgrass, Pre-emergent herbicides must be applied at a critical time. If they are applied to the soil too early, they get washed too deep into the soil or washed away by rainwater. If they are applied too late, the key enzyme inhibited is no longer active. The best control requires a second application about 6–8 weeks later. This provides coverage in mid-late summer when crabgrass can still germinate. Depending on location, one rule of thumb is to apply when the local forsythia blooms are wilting. In the northeast, they need to be applied before azaleas bloom.

Contact Herbicide

Contact herbicides are chemicals that are sprayed onto the crop to kill the already present weed upon contact with minimal damage to the main crop, whereas systemic herbicides are absorbed through the root system. Grapevines are extremely sensitive to the application of contact herbicides which can injury the vine, affect future growth and permanently damage the leaf formations. It is important to avoid spraying contact herbicides during the early growing period in March – the bloom period in June, but spraying should be avoided until October.

Contact Herbicides kill weeds when sprayed directly to the plant parts and typically have a phenoxy-type (ether based) active ingredient. These types of contact herbicides are most widely used on agriculture crops, but grapevines are very susceptible to being damaged from this type of ingredient. Instead, a salt-based active ingredients are preferred for grapevines, or brands like Everest Union, Maverick and Achieve Avenge.

When spraying grapevines, it is very important to reduce chemical drifting by avoiding spraying during the vulnerable period, being aware of wind patterns or high pressure systems, environmental factors (like runoff) and water addition to make the chemical spray heavier and less likely to spread to other areas. Grapevines can also be exposed to drift from other farms or agricultural applications from miles away, so it is crucial to communicate with other agriculturalists when spraying contact herbicides. Most contact herbicides do not require a license to use, but home spraying without the consultation of a professional should be avoided, as these chemicals must be used with caution.

WEED PREVENTION

Prevention is the most effective method of dealing with weeds. Once a weed has entered an area and become established, eradication is far more expensive and it is likely that greater resources will be required to control its further spread and reduce its impact.

The first step in weed prevention, and the most cost effective means of managing weeds.

Once a weed has entered, early detection and eradication is crucial to reduce its potential environmental and economic impacts. It is much easier to treat weeds when present in small numbers than when they are well established.

Early detection and eradication requires an awareness and understanding of the factors that favour the establishment and spread of weeds, and applying appropriate management practices that can prevent or reduce the risks.

The importance of weed spread prevention has grown with the recognition that the spread of most weeds occurs through similar pathways, such as the movement of goods, animals and vehicles contaminated with weed seeds.

Weed Prevention in Agriculture

In agriculture, the pathways for spread include transported livestock and fodder, contaminated crop and pasture seeds, deliberate introductions of new species, and contaminated machinery such as harvesters and recreational vehicles (including boats which can spread water weeds).

There are many ways to prevent weeds in agricultural activities which are well known including:

Restricting the opportunity for new weeds to invade and spread:

- Be vigilant about introducing stock, fodder or seed onto your property to ensure weeds will not be introduced.
- When buying stock, find out where the stock has come from and what weeds infest that area. Buy certified weed free fodder and seed where possible.
- Restrict the movement of vehicles and machinery on your property in periods when seeds are likely to spread.
- Establish tracks and laneways along which vehicle movement can be concentrated.
- Wash down vehicles which have been in known infested areas.
- Do not allow machinery or vehicles to enter your property unless they are clean.

Restricting the spread of existing weed infestations:

- Carry out control works prior to other works.
- Slash and cultivate when weeds are outside of seeding period.
- Work the clean area first and the infested area last. Work from the outside in and clean down equipment prior to moving into a clean area.

Quarantine:

- Hold livestock that may be infested with seed in a single location until they are shorn or until weed seeds have had the chance to pass through their digestive system.
- Feed out infested fodder in a feed lot type situation only and introduce clean fodder to stock.

Monitor:

- Continually monitor weed infestations and carry out control works.

Weed Prevention in your Backyard

Plants from commercial nurseries, landscaping suppliers and gardening clubs can also be pathways for the introduction and spread of weeds. Another significant cause of weed spread is inappropriate use and disposal of garden waste.

There are a large number of potential weeds in gardens. Private gardens contain over 4000 plant species with weed potential, while botanic gardens hold approximately 5000 species of plants with weed potential. The likelihood that any particular plant will become a weed is difficult to predict.

Measures for weed prevention in your backyard include:

- Choose plants that are unlikely to become weeds in your area.

- Check existing garden plants are safe.

- Remove potentially weedy plants.

- Dispose of garden waste carefully.

- Be careful not to spread weeds.

- Place mulch on soil surfaces in the garden to reduce weeds growth.

Weed Prevention in the Natural Environment

Landscapes that contain a diversity of healthy, vigorous vegetation with very little bare ground have the ability, in most cases, to deter weed invasion. It is important to reduce the risk of the environment becoming vulnerable to invasion by exotic species by encouraging beneficial vegetation growth and by avoiding disturbance as much as possible.

Measures for weed prevention in the landscape include:

- Minimise the disturbance of desirable plants along trails, roads, and waterways.

- Maintain desired plant communities through good management.

- Monitor high-risk areas such as transportation corridors and bare ground.

- Revegetate disturbed sites with desired plants.

AQUATIC WEED HARVESTER

Aquatic Plant Harvesters offer an environmentally sound method of controlling excessive aquatic plant growth and nuisance vegetation in waterways of all sizes. These heavy duty work boats are highly efficient in the management of submerged, emergent and free floating aquatic vegetation.

Like an underwater lawn mower, an aquatic weed harvester cuts the vegetation, collecting and storing the weeds on board. Aquatic weed harvesters are fitted with a pick up conveyor at the forward end of the machine, which can be lowered up to six feet deep to cut weeds. One horizontal and two vertical cutter bars sever the vegetation as the machine moves forward through the water, and, when the storage hold becomes full, the weed harvester returns to shore to unload.

Depending on the size of your lake, the type of your weeds, the percentage of weeds you would like to remove, and how quickly you would like to harvest them, we will work with you to design a fully customizable weed harvester to meet your exact needs.

Weed Harvesting Equipment

Weed cutting boats are developed to enable the maintenance of canals, lakes and rivers and to remove excessive aquatic life such as algae and other plants that may negatively affect a waterway's ecology. Mechanical harvesters are large floating machines that have underwater cutting blades that sever the stems of underwater plants, gather the weeds and raise them on conveyor belts, storing the vegetation on board in a hold. Periodically this is discharged to a barge or an onshore facility. The harvested product can be composted, sent to a landfill site or used in land reclamation. In developing countries aquatic vegetation may be harvested by hand or by net from the shore, cut and harvested by boat and lifted ashore by hand, crane, pump or conveyor system. The harvested vegetation may be used for the feeding of livestock. To reduce the high moisture content and to make it easier to transport, the weed can be chopped and pressed. Other uses to which the harvested vegetation can be put include ensiling the material for livestock fodder, adding it to the soil as a bulky organic fertilizer, manufacturing the raw material into pulp, paper or fibre, and fermenting it to produce methane for energy production.

Advantages and Disadvantages

Mechanical harvesters can be effective at clearing aquatic weeds but the machines are expensive and the process may need to be repeated several times in a growing season. Small fragments of weed remain in the water and may spread to other locations thereby aiding in the dispersal of invasive species. Some areas may be too shallow for the mechanical harvester and it may be unable to access restricted locations. Submerged tree stumps can damage the machine. Aquatic weeds can also be utilized as a source of biofuel. An alternative to mechanical harvesting is the use of herbicides, which are easy to apply and less expensive, but may have unwanted adverse impacts on the environment.

HOE

A hoe is an ancient and versatile agricultural and horticultural hand tool used to shape soil, remove weeds, clear soil, and harvest root crops. Shaping the soil includes piling soil around the base of plants (hilling), digging narrow furrows (drills) and shallow trenches for planting seeds or bulbs. Weeding with a hoe includes agitating the surface of the soil or cutting foliage from roots, and clearing soil of old roots and crop residues. Hoes for digging and moving soil are used to harvest root crops such as potatoes.

A farmer using a hoe to keep weeds down in a vegetable garden.

Types

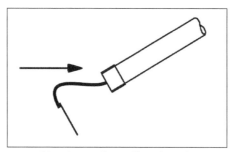
Cultivating tool, a pull or draw hoe.

There are many kinds of hoes of varied appearances and purposes. Some have multiple functions while others have singular and specific functionality.

here are two general types of hoe: draw hoes for shaping soil and scuffle hoes for weeding and aerating soil.

A draw hoe has a blade set at approximately a right angle to the shaft. The user chops into the ground and then pulls (draws) the blade towards them. Altering the angle of the handle can cause the hoe to dig deeper or more shallowly as the hoe is pulled. A draw hoe can easily be used to cultivate soil to a depth of several inches. A typical design of draw hoe, the "eye hoe", has a ring in the head through which the handle is fitted. This design has been used since Roman times.

A scuffle hoe is used to scrape the surface of the soil, loosen the top inch or so, and to cut the roots of, remove, and disrupt the growth of weeds efficiently. These are primarily of two different designs: the Dutch hoe and the hoop hoe.

The term "hand hoe" most commonly refers to any type of light-weight, short-handled hoe, although it may be used simply to contrast hand-held tools against animal or machine pulled tools.

Draw Hoes

Eye hoe heads, some with sow-tooth.

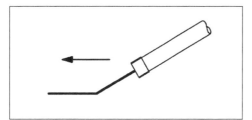

Cultivating tool, a push or thrust hoe.

Hoedad (tree-planting tool).

• The typical farming and gardening hoe with a heavy, broad blade and a straight edge is known as the Italian hoe, grub hoe, grab hoe, pattern hoe, Azada, or dago hoe ("dago" is an ethnic slur referring to Italians, Spaniards, or Portuguese).

• The ridging hoe, also known as the Warren hoe and the drill hoe, is a triangular (point-down) or heart-shaped draw hoe that is particularly useful for digging narrow furrows ("drills") and shallow trenches for the planting of seeds or bulbs.

• The Paxton hoe is similar to the Italian hoe, but with a more rounded rectangular blade.

• The flower hoe has a very small blade, rendering it useful for light weeding and aerating around growing plants, so as not to disturb their shallow roots while removing weeds beyond the reach of the gardener's arm.

- The hoedad, also denominated the hoedag or hodag, is a hoe-like tool used to plant trees. According to Hartzell, "The hoedag was originally called skindvic hoe. Hans Rasmussen, legendary contractor and timber farm owner, is credited with having invented the curved, convex, round-nosed hoedag blade which is widely used today" (emphasis added).

- The mortar hoe is a tool specific to the manual mixing of mortar and concrete, and has the appearance of a typical square-bladed draw hoe with the addition of large holes in the blade.

Scuffle Hoes

- The Dutch hoe is designed to be pushed or pulled through the soil to cut the roots of weeds just under the surface. A Dutch hoe has a blade "sharp on every side so as to cut either forwards and backwards". The blade must be set in a plane slightly upwardly inclined in relation to the dual axis of the shaft. The user pushes the handle to move the blade forward, forcing it below the surface of the soil and maintaining it at a shallow depth by altering the angle of the handle while pushing. A scuffle hoe can easily cultivate soil and remove weeds from the surface layer.

- The hoop hoe, also known as the "action hoe", oscillating, hula, stirrup, pendulum weeder, or "swivel hoe") has a double-edge blade that bends around to form a rectangle attached to the shaft. Weeds are cut just below the surface of the soil as the blade is pushed and pulled. The back and forth motion is highly effective at cutting weeds in loose or friable soil. The width of the blade typically ranges between 3-7 inches. The head is a loop of flat, sharpened strap metal. However, it is not as efficient as a draw hoe for moving soil.

- The collinear hoe or collineal hoe has a narrow, razor-sharp blade which is used to slice the roots of weeds by skimming it just under the surface of the soil with a sweeping motion; it is unsuitable for tasks like soil moving and chopping. It was designed by Eliot Coleman in the late 1980s.

- The swoe hoe is a modern, one-sided cutting hoe, being a variant of the Dutch hoe.

Other Hoes

Fork-hoe.

Hoes resembling neither draw nor scuffle hoes include:

- Wheel hoes are, as the name suggests, a hoe or pair of hoes attached to one or more wheels. The hoes are frequently interchangeable with other tools.

- Horse hoes, resembling small ploughs, were a favourite implement of agricultural pioneer Jethro Tull, claiming in his book "Horse Hoeing Husbandry" that "the horse-hoe will, in wide intervals, give wheat throughout all the stages of its life, as much nourishment as the discreet hoer pleases". The modern view is that, rather than nutrients being released, the crop simply benefits from the removal of competing plants. The introduction of the horse hoe, together with the better-known seed drill, brought about the great increase farming productivity seen during the British Agricultural Revolution.

- Fork hoes, (also known as prong hoes, tined hoes or bent forks) are hoes that have two or more tines at right angles to the shaft. Their use is typically to loosen the soil, prior to planting or sowing.

- Clam hoes, made for clam digging.

- Adze hoes, with the basic hoe shape but heavier and stronger and with traditional uses in trail making.

- Pacul or cangkul (hoes similar to adze hoe from Malaysia and Indonesia).

- Gang hoes for powered use (in use at least from 1887 to 1964).

STALE SEED BED

A false or stale seed bed is a seedbed created some weeks before seed is due to be sown. The early seedbed is used a weed control technique. It is designed to germinate weed seeds that have been disturbed and brought to the soil surface during cultivation, so that the young weeds can then be eliminated. The tilled soil increases the chance of weed seed germination as the fine soil allows seed to grow rapidly than in compacted soil and dormant seeds are brought to the surface. The weeds must then be destroyed before they can create new seeds. By destroying them early, the farmer eliminates most of that season's annual weeds, and nominally increases soil nutrient content.

A stale seed bed technique of weed control creating a seedbed some weeks before seed is due to be sown. The early seedbed is designed to germinate weed seeds that have been disturbed and brought to the soil surface during cultivation, so that the young weeds can then be eliminated before they can propagate.

Method

The technique can be utilized in early spring, when the weather is still too cold for proper seed germination. Several passes are made with a power harrow, such as an R2 Rinaldi, rototiller or plow, then weed seeds are allowed to germinate as weather permits. By tilling, the farmer increases the chance of weed seed germination by the same method as one would for favorable vegetable/ crops: the fine soil allows weed seed to grow rapidly by allowing the seed to open and the roots to spread easier than in compacted soil. Deep tilling will also bring dormant seed to the surface for germination; some species of plant are known for seeds that can lay deeply buried in the soil for years before favorable conditions allow germination.

After weeds have sprouted, they are hoed off or eliminated with the other means (e.g., use of a flame weeder) before sowing of the actual crop. Timing is important; weed seeds must be destroyed before they themselves can create new seeds. By destroying them early, the farmer eliminates most of that season's annual weeds. Turning the dead weeds back into the soil also increases soil nutrient content, although this difference is slight.

In many cases, several tillings are done, perhaps every two weeks beginning in very early spring. This allows more and more weed seeds to germinate only to be killed off later. This eliminates more weeds, but care must be used to not delay planting of a desirable crop later than the crop needs for a successful season's growth. After several years, most, if not all, weeds can be eliminated from the seed bank in the soil. In some cases the effect can be noticed in the same year the process is first carried out.

If the weed patch is vigorous enough, the farmer may mow the patch first before tilling. This allows for easier/quicker decomposition in the soil when the plants are turned under. Some farmers may apply a light and inexpensive fertilizer mix to the soil to hoping to cause even more weed seeds to germinate and eliminate seeds earlier that otherwise would have sprouted in later years.

SOIL STEAM STERILIZATION

Soil steam sterilization (soil steaming) is a farming technique that sterilizes soil with steam in open fields or greenhouses. Pests of plant cultures such as weeds, bacteria, fungi and viruses are killed through induced hot steam which causes vital cellular proteins to unfold. Biologically, the method is considered a partial disinfection. Important heat-resistant, spore-forming bacteria can survive and revitalize the soil after cooling down. Soil fatigue can be cured through the release of nutritive substances blocked within the soil. Steaming leads to a better starting position, quicker growth and strengthened resistance against plant disease and pests. Today, the application of hot steam is considered the best and most effective way to disinfect sick soil, potting soil and compost. It is being used as an alternative to bromomethane, whose production and use was curtailed by the Montreal Protocol. "Steam effectively kills pathogens by heating the soil to levels that cause protein coagulation or enzyme inactivation".

Sheet steaming with a MSD/moeschle steam boiler.

Benefits of Soil Steaming

Soil sterilization provides secure and quick relief of soils from substances and organisms harmful to plants such as:

- Bacteria,
- Viruses,
- Fungi,
- Nematodes,
- Other Pests.

Further positive effects are:

- All weed and weed seeds are killed.
- Significant increase of crop yields.
- Relief from soil fatigue through activation of chemical – biological reactions.
- Blocked nutritive substances in the soil are tapped and made available for plants.
- Alternative to methyl bromide and other critical chemicals in agriculture.

Steaming with Superheated Steam

Through modern steaming methods with superheated steam at 180–200 °C, an optimal soil disinfection can be achieved. Soil only absorbs a small amount of humidity. Micro organisms become active once the soil has cooled down. This creates an optimal environment for instant tillage with seedlings and seeds. Additionally the method of integrated steaming can promote a target-oriented resettlement of steamed soil with beneficial organisms. In the process, the soil is first freed from all organisms and then revitalized and microbiologically buffered through the injection of a soil activator based on compost which contains a natural mixture of favorable microorganisms (e.g. *Bacillus subtilis*, etc.).

Different types of such steam application are also available in practice, including substrate steaming and surface steaming.

Surface Steaming

Several methods for surface steaming are in use amongst which are: area sheet steaming, the steaming hood, the steaming harrow, the steaming plough and vacuum steaming with drainage pipes or mobile pipe systems.

In order to pick the most suitable steaming method, certain factors have to be considered such as soil structure, plant culture and area performance. At present, more advanced methods are being developed, such as sandwich steaming or partially integrated sandwich steaming in order to minimize energy consumption and associated costs as much as possible.

Sheet Steaming

Surface steaming with special sheets (sheet steaming) is a method which has been established for decades in order to steam large areas reaching from 15 to 400 m² in one step. If properly applied, sheet steaming is simple and highly economic. The usage of heat resistant, non-decomposing insulation fleece saves up to 50% energy, reduces the steaming time significantly and improves penetration. Single working step areas up to 400 m² can be steamed in 4–5 hours down to 25–30 cm depth / 90 °C. The usage of heat resistant and non-decomposing synthetic insulation fleece, 5 mm thick, 500 gr / m², can reduce steaming time by about 30%. Through a steam injector or a perforated pipe, steam is injected underneath the sheet after it has been laid out and weighted with sand sacks.

The area performance in one working step depends on the capacity of the steam generator (e.g. steam boiler):

Steam capacity kg/h:	100	250	300	400	550	800	1000	1350	2000
Area m²:	15–20	30–50	50–65	60–90	80–120	130–180	180–220	220–270	300–400

The steaming time depends on soil structure as well as outside temperature and amounts to 1–1.5 hours per 10 cm steaming depth. Hereby the soil reaches a temperature of about 85 °C. Milling for soil loosening is not recommended since soil structure may become too fine which reduces its penetrability for steam. The usage of spading machines is ideal for soil loosening. The best results can be achieved if the soil is cloddy at greater depth and granulated at lesser depth.

In practice, working with at least two sheets simultaneously has proven to be highly effective. While one sheet is used for steaming the other one is prepared for steam injection, therefore unnecessary steaming recesses are avoided.

Depth Steaming with Vacuum

Steaming with vacuum which is induced through a mobile or fixed installed pipe system in the depth of the area to be steamed, is the method that reaches the best penetration. Despite high capital cost, the fixed installation of drainage systems is reasonable for intensively used areas since steaming depths of up to 80 cm can be achieved.

In contrast to fixed installed drainage systems, pipes in mobile suction systems are on the surface. A central suction pipeline consisting of zinc-coated, fast-coupling pipes are connected in a regular spacing of 1.50 m and the ends of the hoses are pushed into the soil to the desired depth with a special tool.

The steaming area is covered with a special steaming sheet and weighted all around as with sheet steaming. The steam is injected underneath the sheet through an injector and protection tunnel. While with short areas up to 30 m length steam is frontally injected, with longer areas steam is induced in the middle of the sheet using a T-connection branching out to both sides. As soon as the sheet is inflated to approximately 1 m by the steam pressure, the suction turbine is switched on. First, the air in the soil is removed via the suction hoses. A partial vacuum is formed and the steam is pulled downward.

During the final phase, when the required steaming depth has been reached, the ventilator runs non-stop and surplus steam is blown out. To ensure that this surplus steam is not lost, it is fed back under the sheet.

As with all other steaming systems, a post-steaming period of approximately 20–30 minutes is required. Steaming time is approximately 1 hour per 10 cm steaming depth. The steam requirement is approximately 7–8 kg/m².

The most important requirement, as with all steaming systems, is that the soil is well loosened before steaming, to ensure optimal penetration.

Negative Pressure Technique

Negative pressure technique generates appropriate soil temperature at a 60 cm depth and complete control of nematodes, fungi and weeds is achieved. In this technique, the steam is introduced under the steaming sheath and forced to enter the soil profile by a negative pressure. The negative pressure is created by a fan that sucks the air out of the soil through buried perforated polypropylene pipes. This system requires a permanent installation of perforated pipes into the soil, at a depth of at least 60 cm to be protected from plough.

Steaming with Hoods

Half automatic steaming hood with three wings in greenhouse.

A steaming hood is a mobile device consisting of corrosion-resistant materials such as aluminum, which is put down onto the area to be steamed. In contrast to sheet steaming, cost-intensive working steps such as laying out and weighting the sheets don't occur, however the area steamed per working step is smaller in accordance to the size of the hood.

Outdoors, a hood is positioned either manually or via tractor with a special pre-stressed 4 point suspension arm. Steaming time amounts to 30 min for a penetration down to 25 cm depth. Hereby a temperature of 90 °C can be reached. In large stable glasshouses, the hoods are attached to tracks. They are lifted and moved by pneumatic cylinders. Small and medium-sized hoods up to 12 m² are lifted manually using a tipping lever or moved electrically with special winches.

Combined Surface and Depth Injection of Steam

Sandwich steaming, which was developed in a project among DEIAFA, University of Turin and Ferrari Costruzioni Meccaniche, represents a combination of depth and surface steaming, offers an efficient method to induce hot steam into the soil. The steam is simultaneously pushed into the soil from the surface and from the depth. For this purpose, the area, which must be equipped with a deep steaming injection system, is covered with a steaming hood. The steam enters the soil from the top and the bottom at the same time. Sheets are not suitable, since a high pressure up to 30 mm water column arises underneath the cover.

Sandwich steaming machine model.

Sandwich steaming offers several advantages. On the one hand, application of energy can be increased to up to 120 kg steam per m²/h. In comparison to other steaming methods up to 30% energy savings can be achieved and the usage of fuel (e.g. heating oil) accordingly decreases. The increased application of energy leads to a quick heating of the soil which reduces the loss of heat. On the other hand, only half of the regular steaming time is needed.

Comparison of sandwich steaming with other steam injection methods relating to steam output and energy demand(*):

Steaming method	Max. steam output	Energy demand (*)
Sheet steaming	6 kg/m²h	about 100 kg steam/m³
Depth steaming (Sheet + vacuum)	14 kg/m²h	about 120 kg steam/m³
Hood steaming (Alu)	30 kg/m²h	about 80 kg steam/m³
Hood steaming (Steel)	50 kg/m²h	about 75 kg steam/m³
Sandwich steaming	120 kg/m²h	about 60 kg steam/m³

(*) in soil max 30% moisture

Clearly, Sandwich steaming reaches the highest steam output at the lowest energy demand.

Partially Integrated Sandwich Steaming

The partial integrated sandwich steaming is an advanced combined method for steaming merely the areas which shall be planted and purposely leaving out those areas which shall not be used. In order to avoid risk of re-infection of steamed areas with pest from unsteamed areas, beneficial organisms can directly be injected into the hygenized soil via a soil activator (e.g. special compost). The partial sandwich steaming unlocks further potential savings in the steaming process.

Container/Stack Steaming

Stack steaming is used when thermically treating compost and substrates such as turf. Depending on the amount, the material to be steamed is piled up to 70 cm height in steaming boxes or in small dump trailers. Steam is evenly injected via manifolds. For huge amounts, steaming containers and soil boxes are used which are equipped with suction systems to improve steaming results. Midget amounts can be steamed in special small steaming devices.

The amount of soil steamed should be tuned in a way that steaming time amounts to at most 1.5 h in order to avoid large quantities of condensed water in the bottom layers of the soil.

Steam Output kg/h:	100	250	300	400	550	800	1000	1350	2000
m³/h about:	1.0–1.5	2.5–3.0	3.0–3.5	4.0–5.0	5.5–7.0	8.0–10.0	10.0–13.0	14.0–18.0	20.0–25.0

In light substrates, such as turf, the performance per hour is significantly higher.

USING IRRIGATION TO MANAGE WEEDS

Surface, sprinkler, and drip irrigation are the three primary types of irrigation methods used to grow crops. Within each method, there are several subcategories, each of which varies in water use efficiency, cost, yield, and weed management potential.

An irrigation canal for furrow irrigation of cabbage (Brassica oleracea) (left), solid set sprinkler irrigation of onion (Allium cepa) (center) and surface drip irrigation of recently seeded cabbage (right).

Surface Irrigation and the Impact on Weeds

Surface irrigation, which floods entire fields or supplies water in furrows between planted rows, is the most common type of irrigation used worldwide. Some surface irrigation systems have been operating continuously for thousands of years and have the ability to supply enormous quantities of water over widespread areas. Flood and furrow irrigation can have water use efficiencies per unit of yield ranging from 25-50% of well managed drip irrigation systems. One of the most common crops grown worldwide with flood irrigation is lowland rice (*Oryza sativa*). Flood irrigation can be an integral part of weed management for this crop.

As a semi-aquatic crop, lowland rice production utilizes substantial quantities of water. It was estimated that more than 2m of water are used per crop of rice grown. This underscores the substantial water requirements for lowland rice production; particularly in the initial flooding stages when large quantities of water may be lost prior to saturation. Although it has been reported that rice grown under saturated field conditions did not experience additional water stress and yielded no differently than rice grown under standing water; rice which is grown under standing water competes better with weeds than when grown in saturated soils. Although some weeds propagate vegetatively, most develop from seeds; thus flooding can restrict the germination and reduce the abundance of many weeds found in rice paddies.

Despite reducing the presence of some weed species, flooded lowland rice fields have over time selected for the presence of semi-or aquatic weed species. To reduce the presence of some of these weeds flooded soils are often tilled. While the primary goal of tillage is to uproot recently germinated weed seedlings; tilling flooded soils can destroy structure and porosity. This results in soils within low infiltration rates, which increases water retention, allowing fields to remain flooded.

Weed control in modern rice production is a system where irrigation management is integrated with tillage and planting practices as well as herbicides. Weed control was better in fields submerged under 20 cm of water compared to those submerged under 5 cm of water when no herbicides were used. However, when herbicides were included weed control improved significantly at all depths. Flowable-granular herbicide formulations, which are often used in lowland rice production, also rely on standing water for dispersal. Flooded paddy fields allow uniform dispersal of low quantities of herbicides resulting in superior control of weeds. The integration of herbicides into the lowland rice production systems has reduced labor requirements for weed control by more than 80% since the introduction of 2,4-D in 1950, while simultaneously improving overall weed management. Flooding has been an effective weed management technique in lowland rice for thousands of years. Coupled with modern herbicides, farmers can efficiently manage weeds on a large scale. Nonetheless, the high costs of water and demands on finite fresh water resources may result in substantial changes to the current lowland rice production system. The development of "aerobic-rice," drought tolerant lowland varieties that can yield well on non-saturated soils, may change how irrigation is used to manage weeds in lowland rice. Aerobic-rice is grown in a manner similar to many other grains, with land allowed to dry between irrigation cycles. This has the potential to reduce the reliance on flooding and irrigation water for weed control, likely shifting to chemical or mechanical methods.

Furrow irrigation is a common irrigation method where water is sent through ditches dug between raised beds to provide water to plants. Instead of flooding entire fields, only furrows between beds are wetted, allowing water to seep into growing beds through capillary action. Furrow irrigation is commonly used on millions of hectares of crops worldwide; where complex canal networks can move irrigation water hundreds of miles from upland sources to lower elevation growing areas. As would be expected, weed pressure in the irrigated furrows between rows is generally higher than with the rows themselves. To control these weeds, mechanical cultivation may be used, but in many instances, herbicides, either applied to the soil as sprays or through irrigation water, are relied upon. However, the administration of herbicides through furrow irrigation can be challenging. Poor application uniformity, downstream pollution, and inaccuracies due to difficulties in measuring large quantities of water are challenges associated with applying herbicides through surface irrigation water. Chemical choice also is important when applying herbicides in surface irrigation systems. For example, large quantities of the herbicide EPTC volatilized shortly after application via flood irrigation in alfalfa (*Medicago sativa*); Variability in the effectiveness of some herbicides when applied through furrow irrigation compared to conventional methods.

A variant on the typical furrow irrigation system has been developed that combines furrow irrigation with polyethylene mulches and rainwater collection to irrigate crops, while controlling weeds. The production method, called the "ridge-furrow-ridge rainwater harvesting system," uses woven, water-permeable, polyethylene mulches that cover two ridges as well as a shallow furrow between the ridges. The system is similar to a raised-bed plastic mulch system, with inter-row areas being left in bare soil. However, unlike a traditional plastic mulch system, a furrow is made in the center of the raised bed to collect any rainwater that ordinarily would be lost as runoff from the bed. This system significantly reduces weed pressure in the furrow area and increases yield with the use of a polyethylene mulch, while reducing the need for supplemental irrigation by collecting rainwater. Interestingly, a similar method of irrigation was employed during early experiments with plastic mulch, prior to the introduction of drip irrigation tubing. In these trials irrigation was achieved by cutting furrows in the soil next to the crop, covering them with plastic, and cutting holes in the plastic for the water to penetrate the plant bed.

Sprinkler Irrigation and the Impact on Weeds

Introduced on a large scale in the 1940s, sprinkler irrigation systems are used on millions of ha of crop land. The three primary types of sprinkler irrigation are center pivot, solid set, and reel or travelling gun systems. Sprinkler systems require a pump to deliver water at high pressures and are costlier than surface irrigation systems, but provide superior application uniformity and require less water to operate. While center pivot systems require relatively level ground; solid set and reel-type systems can be used on with varied topographies. Because of improved application uniformity, sprinkler irrigation is the method of choice when applying herbicides or other agrichemicals through the irrigation system. Sayed and Bedaiwy noted a nearly 8-fold reduction in weed pressure when applying herbicides through sprinkler irrigation compared to traditional methods. Sprinkler irrigation permits growers to uniformly apply water over large areas, which can allow for proper incorporation of some preemergent herbicides. In addition to applying herbicides, preplant sprinkler irrigation of fields, when combined with shallow tillage events after drying, has been shown to significantly reduce weed pressure during the growing season. This process of supplying water to weed seeds prior to planting, which causes them to germinate, where they can then be managed through shallow cultivation or through herbicide application is termed "stale seed-bedding" and is routinely used by farmers in many parts of the US.

Drip Irrigation and the Impact on Weeds

Introduced on a large scale in the late 1960s and early 1970s, drip irrigation has steadily grown in popularity. Although drip irrigation is only utilized on approximately 7% of the total irrigated acreage in the US, it is widely used on high value crops such as berries and vegetables. Drip irrigation, if properly managed, is highly efficient with up to 95% application efficiencies. The productivity of drip irrigation has prompted significant increases (> 500%) in its use over the previous 20-30 years. While drip irrigation is typically expensive and require significant labor to install and manage; the water savings compared to other methods of irrigation have prompted grower adoption. Drip irrigation has several benefits in addition to improved water use efficiencies. By only wetting the soil around plants leaves are kept dry reducing foliar disease and potential for leaf burn when using saline water. Fertilizers, which are easily supplied through drip irrigation, are restricted to an area near active rooting. This leads to more efficient use by the target crop. Because drip irrigation only wets the soil in the vicinity of the drip line or emitter, growers are able to supply irrigation water only in the areas required to grow the crop of interest. Soils between rows are not supplied with water or fertilizer, reducing weed growth. When drip irrigation is coupled with plastic mulch and preplant soil fumigation, weeds can be effectively controlled within rows, leaving only between-row areas to be managed. By restricting weed management to areas between rows growers increase their chemical and mechanical control options. While many farmers may apply preemergent herbicides to between-row areas, weeds that do germinate can be controlled easily with directed sprays of postemergent herbicides with low risk to the crops growing in the plastic mulch. In arid growing regions the combination of plastic mulch and drip irrigation may lead to acceptable weed control without the use of herbicides.

Because drip irrigation can supply limited quantities of water to an area immediately surrounding the crop root zone, it can be ideally suited for insecticide or fungicide injection. The small quantities of water delivered with drip irrigation requires significantly less chemical to maintain a given concentration applied to plants compared to surface or sprinkler irrigation. However, while drip irrigation is one of the most efficient means to deliver chemicals such as systemic insecticides to plants, it is much less effective than comparable sprinkler systems for herbicide applications. The limited wetting pattern and low volume of water used for drip irrigation means that herbicides do not reach much of the cropped area. Within wetted areas herbicides may be degraded prior to the end of the season. Because drip systems are often designed for frequent, low-volume irrigations, soils around plants may remain moist, reducing the efficacy of preemergent herbicides. Fischer reported significantly better weed control when using micro sprinklers compared to drip irrigation in vineyards and orchards. This was due to a reduction in the effectiveness of preemergent herbicides in drip irrigated treatments late in the growing season. The drip irrigated plants had persistent soil moisture near the emitters resulting in enhanced degradation of the applied herbicides. Drip irrigation is often used in tandem with herbicides; however, they are often applied using conventional sprayers. Therefore, the weed control benefits of drip irrigation are due to the ability to precisely manage and locate water where it will most benefit crops while reducing availability for weed growth. One method that allows growers to precisely locate water in the root zone, below the soil surface, away from weed seeds is subsurface drip irrigation.

Subsurface Drip Irrigation

Subsurface drip irrigation (SDI) has been utilized in various forms for more than a century. Presently SDI uses standard drip irrigation tubing that is slightly modified for below-ground use. While typical surface drip irrigation tubing have walls that are usually 8 or 10-mil thick; tubing made specifically for multi-season SDI applications, have walls with a 15-mil thickness. In addition, tubing made specifically for SDI applications may have emitters which are impregnated with herbicides to prevent root intrusion. Because growers are unable to inspect buried tubing, any problems with emitter clogging or cuts in the line may go unnoticed for long periods of time. Subsurface drip irrigation used for the production of high-value crops such as vegetables, which tend to have shallow root systems, may be buried at depths of 15-25 cm. Subsurface drip tubing that is used for agronomic crops such as cotton (*Gossypium spp.*) or corn (*Zea maize*) is generally buried 40-50 cm below the soil surface. Drip irrigation tubing used for agronomic crops is typically left in place for several years in order to be profitable and must reside below the tillage zone to avoid being damaged. Agronomic crops in general tend to be deeper rooted than many vegetable crops allowing them to access water supplied at greater depths. In addition, the deeper placement of the irrigation tubing reduces the potential rodent damage, which can be significant.

Drip tubing may be placed during or after bed formation in tilled fields or into conservation tillage fields with drip tape injection sleds. While SDI that is used for a single season may be connected to flexible "lay-flat" tubing at the ends of fields; more permanent installations are generally coupled to rigid PVC header lines.

Injection sled for SDI.

Although concern over buried drip tubing collapsing under the pressure of the soil above is justified; properly maintained SDI systems have lasted 10-20 years in the Great Plains without significant problems. For permanent systems, lines must be cleaned and flushed after every crop if not more frequently. In single-season trials conducted, end of season flow rates were found to be no different between surface and SDI systems placed at a depth of 15 cm (*T. Coolong, unpublished data*). However, when comparing SDI that had been in use for three years for onion production to new SDI tubing, there were slight reductions in discharge uniformity in the used tapes.

Subsurface Drip Irrigation in Organic Farming

Some of the earliest uses of SDI were not based on enhanced water use efficiency but because drip irrigation tubing on the soil surface could interfere with agricultural equipment, particularly cultivation tools. While many conventional farmers now rely more on chemical weed control than on cultivation, most organic growers must rely exclusively on cultivation to manage weeds. For this reason, SDI is particularly appropriate for organic farming systems. Traditional placement of drip irrigation tubing requires growers to remove the tubing prior to cultivation, increasing labor costs. By burying drip tubing below the depth of cultivation, growers can control weeds mechanically. SDI is routinely used for bare-ground, organic vegetable production at The University of Kentucky Center for Horticulture Research. This system uses a SDI injection sled coupled with in-row cultivators to effectively control weeds in a humid environment.

Buried drip irrigation tubing entering the soil at the end of a field (top left), a two-row cultivator using side-knives and spring hoes (top right), a rolling basket weeder controlling weeds within and between rows (bottom left), and organically-managed kale and collard (Brassica oleracea Acephala group) crops (bottom right) that are grown with SDI and mechanical cultivation for near complete weed control in a humid environment.

In this system, SDI tubing is placed approximately 15 cm below the surface on a shallow raised bed. Using SDI in combination with precision cultivation has allowed for nearly complete control of weeds on an organic farm in an environment which may regularly experience 25 cm or more rain during the growing season.

Subsurface Drip Irrigation and Water Use

More than 40 types of crops have been tested under SDI regimes. In most cases yields with SDI were no different than or exceeded yields for surface drip irrigation. In many cases water savings were substantial. However, SDI relies on capillary movement of water upward to plant roots. Soil hydraulic properties can significantly affect the distribution patterns of water around emitters, making interpretation of data difficult when comparing the effectiveness of SDI in different soil types. Trials often report water savings or increased yield in SDI systems compared to surface drip systems for vegetable crop production although some do not.

In 2012, studies were conducted at The University of Kentucky Center for Horticulture Research (Lexington, KY, US) comparing SDI at a depth of 15 cm to surface placement of drip irrigation tubing for the production of acorn squash (*Cucurbita pepo*) 'Table Queen'. The soil was a Maury silt loam series, mesic Typic Paleudalfs. Irrigation was controlled automatically with switching-tensiometers placed at a depth of 15 cm from soil surface. Tensiometers were placed approximately 20 cm from plants and 15 cm from the drip tubing which was centered on raised beds. Tensiometer set points were as follows: on/off -40/-10 kPa and -60/-10 kPa for both SDI and surface drip systems. In both moisture regimes the surface applied drip irrigation utilized less water during the growing season than SDI. Interestingly, the number of irrigation events and the average duration of each event varied significantly among the surface and SDI treatments when irrigation was initiated at -40 kPa, but were similar when irrigation was scheduled at -60 kPa. Irrigations were frequent, but relatively short for the -40/-10 kPa surface irrigation treatment. Comparable results have been reported in studies conducted in tomato (*Lycopersicon esculentum* syn. *Solanum lycopersicum*) and pepper (*Capsicum annuum*) using a similar management system and set points. However the SDI -40/-10 kPa treatment irrigated relatively infrequently and for longer periods of times. When irrigation was initiated at -60 kPa and terminated at -10 kPa there were differences in water use between the two drip systems, with the surface system being more efficient. However, unlike the -40/-10 kPa treatments, the numbers of irrigation events were not different between the two drip irrigation systems. The difference in the response of the SDI and surface systems when compared under different soil moisture regimes was not expected and suggests that irrigation scheduling as well as soil type may have a significant impact on the relative performance of SDI compared to surface drip irrigation. This should be noted when comparing the performance of SDI and surface drip irrigation systems.

Subsurface Drip Irrigation for Improved Weed Management

A key benefit of SDI is a reduction in soil surface wetting for weed germination and growth. Although the lack of surface wetting can negatively impact direct-seeded crops, transplanted crops often have significant root systems that may be wetted without bringing water to the soil surface. Direct-seeded crops grown with SDI are often germinated using overhead microsprinkler irrigation. The placement of SDI tubing as well as irrigation regime can impact the potential for surface wetting and weed growth. SDI is often located 40-50 cm below the soil surface in most agronomic crops, but is typically shallower (15-25 cm) for vegetable crops.

Irrigation treatment on/off	Irrigation type	Events	Mean irrigation time	Mean irrigation vol.
kPa		no.	min/event	l·ha⁻¹
-40/-10	Surface	48	92	1.25×10^6
-40/-10	SDI	18	276	1.50×10^6
-60/-10	Surface	14	201	0.84×10^6
-60/-10	SDI	14	251	1.06×10^6
Mean number of irrigation events, irrigation time per event, and irrigation volume for the season 'Table Queen' squash grown with automated irrigation in 2012 in Lexington, KY.				

Table: A comparison of SDI and surface drip irrigation under two automated irrigation schedules.

SDI not only keeps the soil surface drier, but also encourages deeper root growth than surface drip systems. In bell pepper, a shallow rooted crop, SDI encouraged a greater proportion of roots at depths below 10 cm when laterals were buried at 20 cm. Encouraging deeper root growth may afford greater drought tolerance in the event of irrigation restrictions during the production season.

In arid climates SDI has been shown to consistently reduce weed pressure in several crops, including cotton, corn, tomato, and pistachio (*Pistacia vera*). For example, weed growth in pistachio orchards in Iran was approximately four-fold higher in surface irrigated plots compared to those with SDI. In humid regions, benefits may depend on the level of rainfall received during the growing season; however, a reduction in the consistent wetting of the soil surface should allow for a reduction in weed pressure, particularly when coupled with preemergent herbicides.

Processing tomatoes represent one of the most common applications of SDI in vegetable crops. The impact of SDI (25 cm below the soil surface) and furrow irrigation on weed growth were compared in tomato. When no herbicides were applied, annual weed biomass was approximately 1.75 and 0.05 tons per acre dry weight in the furrow and SDI treatments, respectively. With herbicides, both irrigation treatments had similar levels of weed biomass. However, in that study, weed biomass in the SDI non-herbicide treatment was similar to the furrow irrigation with herbicide treatment, suggesting that when using SDI, herbicides may not be necessary in arid environments.

A similar trial compared SDI and furrow irrigation across different tillage regimes with and without the presence of herbicides in processing tomato. In that study, both conservation tillage and SDI reduced the weed pressure compared to conventional alternatives. However, when main effects were tested, SDI had the largest impact on weed growth of any treatment. Main effects mean comparisons showed that SDI treatments had weed densities of 0.5 and 0.6 weeds per m² in the planting bed in years one and two of the trial, respectively, compared to 17.9 and 98.6 weeds per m² in the plant bed for furrow irrigated treatments. As would be expected, SDI substantially reduced weed populations in the furrows between beds as they remained dry during the trial. In this trial SDI had a greater impact on weed populations than herbicide applications. SDI could reduce weed populations sufficiently in conservation tillage tomato plantings in arid environments such that herbicides may not be necessary. In another related trial, weed populations were evaluated for processing tomatoes grown with SDI and furrow irrigation under various weed-management and cultivation systems. The relatively small amounts of water used in drip irrigation underscore the need for proper scheduling; otherwise water deficits can occur, resulting in poor yields.

The difference in weed growth approximately 10 days after transplanting between acorn squash (*Cucurbita pepo*) which were subjected to SDI at a depth of 15 cm below the soil surface (left) and surface drip irrigation (right). A preemergent herbicide (halosulfuron methyl, Sandea) was applied to all plots.

Efficient Management of Drip Irrigation

Appropriate management of irrigation requires growers to determine when and how long to irrigate. A properly designed and maintained drip irrigation system has much higher application efficiencies than comparable sprinkler or surface irrigation systems. However, even with drip irrigation, vegetable crops can require large volumes of water - more than 200,000 gallons per acre for mixed vegetable operations in Central Kentucky, US. Poorly managed drip irrigation systems have been shown to reduce yields and waste significant quantities of water. Just 5 h after the initiation of drip irrigation, the wetting front under an emitter may reach 45 cm from the soil surface, effectively below the root zone of many vegetables. When drip irrigation is mismanaged, a key benefit – limiting water available for weeds, is lost. The ability to precisely apply water with drip irrigation means that a very high level of management can be achieved with proper scheduling.

Irrigation scheduling has traditionally been weather or soil-based; although several plant-based scheduling methods have been proposed. In weather-based scheduling, the decision to irrigate relies on the soil-water balance. The water balance technique involves determining changes in soil moisture over time based on estimating evapotranspiration (Et) adjusted with a crop coefficient. These methods take environmental variables such as air temperature, solar radiation, relative humidity and wind into account along with crop coefficients that are adjusted for growth stage and canopy coverage. Irrigating based on Et can be very effective in large acreage, uniformly planted crops such as alfalfa, particularly when local weather data is available. However, irrigating based on crop Et values for the production of vegetable crops is prone to inaccuracies due to variations in microclimates and growing practices. Plastic mulches and variable plant spacing can significantly alter the accuracy of Et estimates. Furthermore the wide variability observed in the growth patterns in different cultivars of the same vegetable crop can substantially alter the value of crop coefficients at a particular growth stage. In many regions of the US, producers do not have access to sufficiently local weather data and the programs necessary to schedule irrigation.

An alternative to using the check-book or Et-based models for irrigation is to use soil moisture-based methods. Perhaps the simplest and most common method is the "feel method," where irrigation is initiated when the soil "feels" dry. Experienced growers may become quite efficient when using this method. More sophisticated methods of scheduling irrigation may use a tensiometer or granular matrix type sensor.

These methods require routine monitoring of sensors, with irrigation decisions made when soil moisture thresholds have been reached. This requires the development of threshold values for various crops and soil types. Soil water potential thresholds for vegetable crops such as tomato and pepper have been developed. Drip irrigation is well suited to this type of management as it is able to frequently irrigate low volumes of water allowing growers to maintain soil moisture at a near constant level. In some soils, high-frequency, short-duration irrigation events can reduce water use while maintaining yields of tomato when compared to a traditionally scheduled high-volume, infrequent irrigation.

Table: A comparison of high frequency short duration to more traditional infrequent but long duration irrigation scheduling using soil moisture tension to schedule irrigation.

Irrigation on/ off	2009			2010		
	Events	Mean irrigation time	Mean irrigation vol.	Events	Mean irrigation time	Mcan irrigation vol.
kPa	no.	min/event	l·ha-1	no.	min/event	l·ha-1
-30/-10	39	110	1.30 x 106	28	144	1.22 x 106
-30/-25	59	91	1.63 x 106	22	140	0.93 x 106
-45/-10	21	221	1.41 x 106	22	167	1.11 x 106
-45/-40	76	40	0.92 x 106	18	146	0.79 x 106

Irrigation delivered frequently for short durations so as to maintain soil moisture levels in a relatively narrow range could save water and maintain yields, but efficiencies varied depending on season and the soil moisture levels that were maintained. In two years of trials, irrigation water was most efficiently applied when soil moisture was maintained between -45 and -40 kPa for tomatoes grown on a Maury Silt Loam soil. However, when soil moisture was maintained slightly wetter at -30 to -25 kPa, thc rclative application efficiency was affected by growing year. Therefore, while an effective method, soil moisture-based irrigation scheduling may produce variable application efficiencies and should be used in concert with other methods.

After more than 40 years of research with drip irrigation, results suggest that a mix of scheduling tactics should be employed to most efficiently manage irrigation. Using a method incorporating climate factors and the water-balance technique, one could increase relative efficiency compared to the baseline by 60-70%. However, when soil moisture sensors were combined with Et-based methods, the relative efficiency of drip irrigation could be increased by more than 115% over a fixed interval method. Therefore multiple strategies should be used to optimize drip irrigation scheduling. This ensures maintaining yields while reducing excessive applications of water, reducing the potential for weed growth.

References

- Weed-management, agriculture-forestry-forest-protection: vikaspedia.in, Retrieved 18 July, 2019

- Cultural-weed-control, pest-management-weeds: knowledgebank.irri.org, Retrieved 16 May, 2019

- Biological-control, weed: wssa.net, Retrieved 14 March, 2019

- Overview-of-biological-methods-of-weed-control, biological-approaches-for-controlling-weeds: intechopen.com, Retrieved 05 January, 2019

- Chemical-control, plant-weed: britannica.com, Retrieved 23 June, 2019

- Contact-herbicide- 1273: winefrog.com, Retrieved 06 July, 2019

- Aquatic-weed-harvesters: aquarius-systems.com, Retrieved 18 February, 2019

- Using-irrigation-to-manage-weeds-a-focus-on-drip-irrigation, weed-and-pest-control-conventional-and-new-challenges: intechopen.com, Retrieved 08 March, 2019

- National Research Council (2002). Making Aquatic Weeds Useful: Some Perspectives for Developing Countries. The Minerva Group. pp. 67–68. ISBN 978-0-89499-180-6

PERMISSIONS

We would like to thank the editorial team for lending their expertise to make the book truly unique. They have played a crucial role in the development of this book. Without their invaluable contributions this book wouldn't have been possible. They have made vital efforts to compile up to date information on the varied aspects of this subject to make this book a valuable addition to the collection of many professionals and students.

This book was conceptualized with the vision of imparting up-to-date and integrated information in this field. To ensure the same, a matchless editorial board was set up. Every individual on the board went through rigorous rounds of assessment to prove their worth. After which they invested a large part of their time researching and compiling the most relevant data for our readers.

The editorial board has been involved in producing this book since its inception. They have spent rigorous hours researching and exploring the diverse topics which have resulted in the successful publishing of this book. They have passed on their knowledge of decades through this book. To expedite this challenging task, the publisher supported the team at every step. A small team of assistant editors was also appointed to further simplify the editing procedure and attain best results for the readers.

Apart from the editorial board, the designing team has also invested a significant amount of their time in understanding the subject and creating the most relevant covers. They scrutinized every image to scout for the most suitable representation of the subject and create an appropriate cover for the book.

The publishing team has been an ardent support to the editorial, designing and production team. Their endless efforts to recruit the best for this project, has resulted in the accomplishment of this book. They are a veteran in the field of academics and their pool of knowledge is as vast as their experience in printing. Their expertise and guidance has proved useful at every step. Their uncompromising quality standards have made this book an exceptional effort. Their encouragement from time to time has been an inspiration for everyone.

The publisher and the editorial board hope that this book will prove to be a valuable piece of knowledge for students, practitioners and scholars across the globe.

INDEX

CPSIA information can be obtained
at www.ICGtesting.com
Printed in the USA
BVHW010145270322
632559BV00003B/73

9 781647 400705